LIVING WITH THE DESERT

This book is dedicated to the villagers of the Plateau who so courteously showed us their buildings and so imaginatively interpreted our questions

LIVING WITH THE DESERT

WORKING BUILDINGS OF THE IRANIAN PLATEAU

Elisabeth Beazley & Michael Harverson

(with a contribution by Susan Roaf)

British Library Cataloguing in Publication Data

Beazley, Elisabeth
Living with the desert.
1. Architecture – Iran
I. Title II. Harverson, Michael
720'.955'35 NA1480

ISBN 0 85668 192 X

Printed and Published by ARIS & PHILLIPS LTD,
Teddington House, Warminster, Wilts, England.

CONTENTS

ACKNOWLEDGMENTS

This book owes a great deal to many friends who have most generously shared their knowledge and enthusiasm with us. We are very grateful to them all. Since we both worked separately in Iran, the list is even longer than it might otherwise have been.

The British Institute of Persian Studies, notably its Assistant Secretary, Mary Gueritz, and its founding Director, David Stronach, have given important help and encouragement over many years, especially in connection with Elisabeth Beazley's fieldwork. Friends at The School of Oriental and African Studies, particularly David Morgan, have tolerated a stream of recondite questions and given wise advice when answers were hard to find. Michael Harverson's fieldwork in 1977 was only made possible by generous funds from the Goldsmiths' Company, and by Keith Turner, who granted sabbatical leave.

Other institutions which have aided our research have been St John's College, Oxford, and the libraries of the Development Planning Unit (University College, London), The India Office, The London School of Hygiene and Tropical Medicine, The Royal Geographical Society, and The School of Oriental and African Studies. We are also grateful to the Editors of *Country Life* and *Iran* for use of articles which originally appeared in these journals. Among authors whom we have never met, but whose books have often been an invaluable source of information and stimulus, we would like to single out Hans Wulff, who recorded much of Iranian technology in the 1930's, and Dame Freya Stark, who travelled 'curiously' and affectionately throughout the Middle East for over half a century.

Martin Watts and Alan Stoyel of the Wind and Watermill Section of the Society for the Protection of Ancient Buildings have frequently made time to consider the apparent vagaries of Iranian corn-milling. Richard Buxton has struggled loyally with some unrewarding photographs. Susan Wright, who worked for several years in Iran on anthropological research, has provided much detailed information, some of which she kindly acquired with us in mind. Ingrid Holford's advice on matters climatical has been invaluable. We are also most grateful to Laura Webb, Ray Harrison, and K. Mathey for their help in various ways.

Susan Roaf, who made a special study of the windcatchers of the Yazd region, generously allowed us to use her work in Chapter 5, for which she also drew the relevant diagrams; only her residence in Baghdad during the last few years has prevented her from playing a fuller part in the making of this book. Valuable help in Iran has been given by Olive Kitson, who took many of the photographs, Rosemary Carlisle, Carol and Robert Hillenbrand, Ralph Pinder-Wilson, Martin Weaver, and the late Dr. Caro Minasian.

We are grateful to the following for permission to reproduce their photographs: John Donat (3, 11, 15, 41), The Fitzwilliam Museum (143, 145), Herb Garcia (144), Robert Harding (7), Richard Hewer (108, 113, 122), T.F. Hewer (127), The India Office Library (59), Olive Kitson (88, 89, 90, 101, 129, 130, 133, 136, 137, 138), Susan Roaf (44, 73, 76, 79), Anthony Smith (29), Martin Weaver (92) and Susan Wright (82, 97, 98).

Adrian Phillips has guided us most patiently in the final stages of preparing the book for publication. We are especially indebted to Michael and Rosemary Halsted who read the whole manuscript and made numerous valuable suggestions, as did David Walters; and to Gill Barrett for her work on much of the typescript.

The mistakes that remain are our own, and we would welcome corrections and comments.

A special tribute is due to Michael's wife, Susan, and to Elisabeth's husband, David, whose support and forbearance during the many months of assembling this book have been remarkable. All too often we have sympathised with the wife of the tenth-century lexicographer al-Jalil, who is supposed to have burnt her husband's dictionary in anger that he spent so much time on it.

Lastly, we should like to record our gratitude to all those Iranians who have helped us, often complete strangers, with hospitality, directions, reminiscences, answers and simple friendship.

December, 1981.

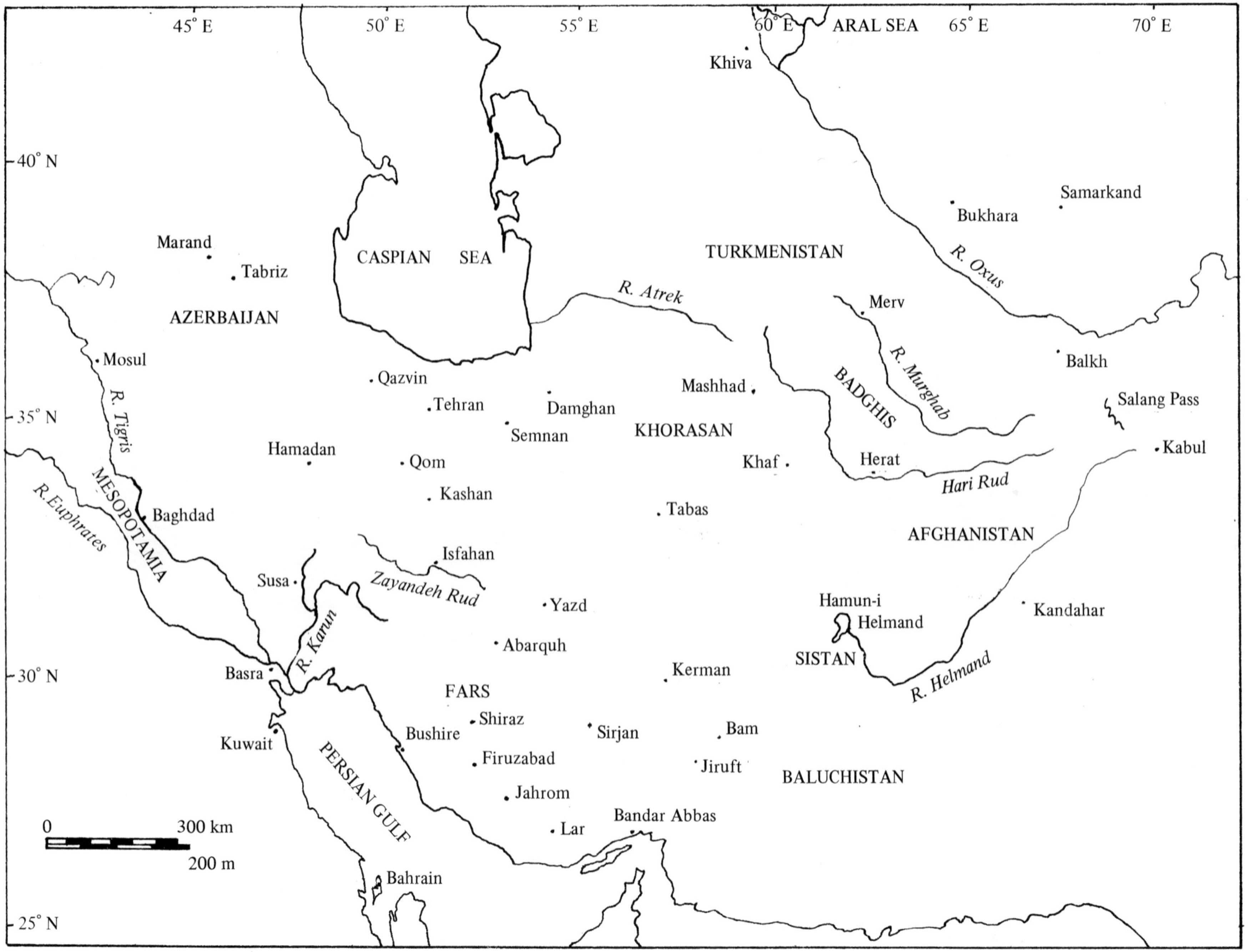

Map 1: Iran and neighbouring areas, with settlements, rivers and regions mentioned in this book.

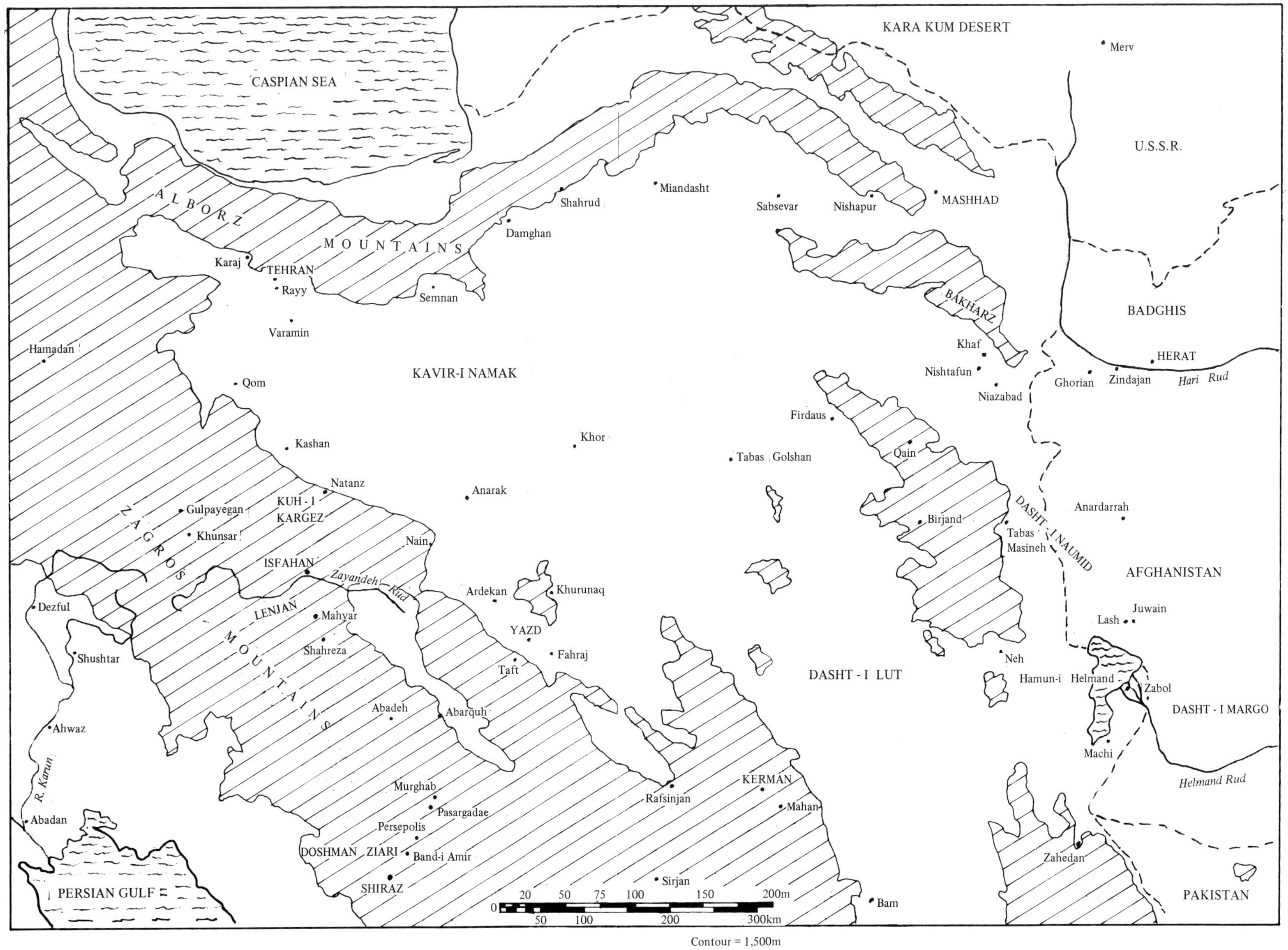

Map 2: The Iranian Plateau

GLOSSARY

āb-anbār	آب انبار	cistern
ābdang	آب دنگ	rice-hulling watermill
aivān	ایوان	large open porch
ājur	آجور	fired brick
arūbah		Hebrew term for a vertical stone penstock
āsīā	آسیا	mill
āsīāb	آسیاب	watermill
āsīā-ye bādī	آسیای بادی	windmill
bād	باد	wind
bādgīr	بادگیر	wind-catcher
bāgh	باغ	garden
band	بند	dam
borj-i kaftar	برج کفتر	pigeon-tower
dasht	دشت	desert (the general word)
gach	گچ	plaster of Paris, gypsum
gel	گل	mud
hammām	حمام	communal bath house
haram serai	حرم سرا	private quarters of a house or palace
havā-kesh	هوا کش	ventilator
hōz	حوض	pool
jūb	جوی	ditch or larger water channel
kāhgel	کاهگل	plaster of mud & straw
kārīz	کاریز	qanat
kaval	کول	baked clay hoop used to strengthen a qanat
kavīr	کویر	mud flats with a salt crust
khābīdeh	خوابیده	asleep, disused
khāneh	خانه	room
khesht	خشت	mud brick
lūt	لوت	barren desert
mādar-chāh	مادر چاه	deepest shaft at a qanat head
maqsam	مقسم	pierced stone slab for water distribution
mihrāb	محراب	prayer niche, orientated towards Mecca
muqanni	مقنی	professional qanat digger
namak	نمک	salt
noria	ناعورة	water-raising wheel : Spanish version of Arabic word
pankeh	پنکه	ceiling fan
qanāt	قنات	subterranean water conduit
sang	سنگ	stone
sāqīyah	ساقیه	endless chain of pots system of water raising
shahr	شهر	city
tālār	تالار	upper reception room raised above the courtyard
vaqf	وقف	money or property administered as a religious charity
yakh-chāl	یخچال	ice-house
yeilāq	ییلاق	(nomadic) summer quarters in the hills

Windmilling terms used in Nishtafun (Chapter 7)

ārjīnah	آرجینه	bill for dressing stones
bāzū	بازو	spoke
damardtā	ده مرد تا	long lever used when raising millstones
dāneh kesh	دانه کش	pegs laid in feeder channel
dar bād	در باد	winnowing inlet
dargāh bālā	درگاه بالا	windslit
dūl	دول	grain hopper
charkh o par	چرخ و پر	windwheel
gūdarān	گودران	stick to support windwheel on roof when raised to permit stone dressing
kalleh tīr	کله تیر	top bearing
kharpol	خرپل	crossbeam
laklakā	لکلکا	vibrator stick
mīkh	میخ	thrust pin
mosht	مشت	lightening device for altering gap between millstones
nay	نای	reed mat (sail itself or drawn across windslit)
nāvdān	ناودان	feeder channel
par	پر	sail
peiwand	پیوند	detachable bottom section of windshaft
sang-i bālā	سنگ بالا	runner stone
sang-i pāīn	سنگ پاین	bedstone
takht	تخت	footbeam
tīr	تیر	windshaft
tobreh	تبره	rynd

INTRODUCTION

This book is about living *with* the desert, not *in* it. It deals with the towns and villages on the desert fringe and in its oases, rather than desert travellers or the nomad way of life. It might be said to be analogous to a book about ports and harbours and islands rather than about the ships and seamen who cross the sea.

Much has been written about Persian architecture, ranging from mosques and palaces to shrines and garden pavilions. The subject here is the vernacular tradition as seen in the context of settled life on the Iranian Plateau, a largely desert area with extremes of climate.

Our surveys were made gradually between 1961 and 1977. The first attraction of many of the buildings was architectural: the sculptural form of domed and vaulted villages, the fretted sky-line of a town ventilated by wind catchers (fig. 81). Then followed the fascination of trying to discover how some of the buildings worked: ice-houses in the desert, the efficiency of windmills with crude sails set on vertical shafts to utilise particular winds. Although the importance of water to all life on the Plateau was immediately evident, it took time to understand how it governed not only the pattern of settlement but also so many aspects of daily life. Suddenly the energy crisis gave these buildings topical interest, and we found new allies in our search.

Before the recent political upheaval put an end to our field-work, we felt that enough material had been gathered to make some contribution to this subject, to which no previous book and only a few scattered articles have been devoted. The Plateau is vast, so there must be important exceptions and additions to all our findings. We realise how incomplete our contribution is, but we hope it will encourage the exploration of other vernacular buildings. The bazaar, with its shops and courtyards, and the *hammam* (bath house), for instance, important features of Persian life, have been reluctantly omitted. Indeed, the houses themselves demand a whole book; and caravanserais another. At least the general aspects common to all mud buildings on the Plateau apply to them equally.

As will be seen from the maps, we have referred at times to vernacular buildings on the Plateau some way from the desert proper, as defined in Chapter 1. The Plateau contains the desert, but also includes large, semi-arid areas where the same building traditions are followed. Again, a consideration of watermills in a region where water must be conjured so ingeniously from the ground naturally entails reference to similar or different types elsewhere in the Middle East. The same holds true in our discussion of wind-catchers and ice-houses.

The vernacular building of arid countries where mud is the chief building material has much in common, so we hope the book may be of interest to those who are working in similar conditions in other parts of the world. For those who know and love Iran the book, we feel, will certainly be nostalgic.

Sadly, some of the buildings which we recorded were already becoming ruins and may by now have disappeared: the earth of their walls once more indistinguishable from the earth of the Plateau. Unlike stone, mud walls need constant maintenance if they are to survive, and naturally enough, once a building is no longer needed maintenance ceases. Oil has superseded the wind and water power of the mills, modern fertilisers have supplanted pigeon dung, and refrigerators have taken the place of ice-houses. Nevertheless, we still make a plea for the conservation of a number of these traditional examples of Persian inventiveness and architectural skill. Soon it will be too late.

As might be expected we know little of what has happened on the Plateau since 1978. This has made it difficult at times to choose the right tense for a verb when writing the book. But this is a very small matter when it is considered in its context and time-scale. It is thousands of years, not merely centuries, since the Persians began to develop their remarkable understanding for living where there is little water; and to demonstrate their rare genius for making life not merely possible but civilized under harsh conditions.

1 THE PLATEAU

2. A spring-fed oasis in the country south-west of the road between Yazd and Kerman where the Plateau begins to rise from the desert.

The heart of Iran is a vast desert, covering many thousands of square miles at a height of between 500 and 1500m. (c.1600–5000ft) above sea level. This desert is crossed by few tracks, and is shunned even by the nomads, who are adept at making life out of very little. Its medieval Arabic name was *mafazah* (wilderness), and its desolation calls to mind "the waste howling wilderness" in the Authorised Version of the Bible (Deuteronomy 32, 10). It is one of the driest and, in summer, one ot the hottest places on the earth's surface.

Yet on the edge of this desert most of the great cities of Iran have flourished, from Siyalk c.4000 B.C. to modern Tehran, asserting a creative strength in the face of its harsh hostility. Man has chosen to remain here, living with the desert on his doorstep. He has managed to come to terms with the ferocity of the sun's heat, discovering over 2,500 years ago how to exploit the water resources deep below the surface, and developing a simple style of building notable for its effective response to the climate and its harmony with its surroundings. He gives the lie to the desert when it seems to deny any chance of survival, let alone dignity, to beings foolhardy enough to settle on its borders.

The desert and its raised rim constitute the Iranian Plateau. It is bounded on the north by the Alborz range, on the west by the Zagros mountains, and on the south by the wide belt of mountains that shut off Kerman from the Persian Gulf and the Gulf of Oman. So the Plateau looks east across the less dramatic highlands near the modern boundary of Iran to the bleak, ominously named Dasht-i Naumid and Dasht-i Margo (deserts of hopelessness and of death) in Afghanistan.

Most travellers between Central Asia and the Near East have had to endure the rigours of the Plateau. The merchants on the Silk Road and the Mongol hordes alike followed in the footsteps of the Aryan Iranians who, early in the second millenium B.C., entered the Plateau at its north-eastern corner and moved westward into Iran along the northern fringe of her central desert.

The strong historical links between the Plateau and the similar country in western Afghanistan and Turkmenistan have been respected in this book. Herat, Balkh and Merv were important cities of Khorasan, the greatest province of medieval Iran, and Persian is still widely spoken throughout the region between the Hari Rud and the Oxus. In the nineteenth century Turkmen raids and Russian incursions south over the Oxus, together with moves in "the Great Game" involving Afghanistan, broke the political unity of Khorasan; but the way of life of the indigenous peasants remained unchanged, responsive to the same sun, wind and weather.

Settlement on the Plateau

One marvel of the Plateau is that men live there. The nomads have kept mainly to the mountains round the edge, whilst maintaining close relations with the towns below them. Except from the north-east, little human traffic has entered the Plateau in order to settle, chancing their luck within the arms of the horseshoe represented by the Alborz and the Zagros. They have joined the descendants of the earliest settlers, who acquired certain survival skills when the Plateau was covered with savannah rather than salt-flats. To press further west would have meant negotiating high mountains, dodging nomads and brigands and then facing likely rejection at the hands of the already adequate, settled population of Mesopotamia. Immigrants from the north-east have not sought the sea or the enervating climate of the date groves beside the Tigris. They have been content to share or conquer the settlements on the Plateau, which seemed to have come to terms with the forbidding environment. The tradition of living under such conditions is so old that its acceptance is second nature, much older than the Muslim resignation to God's inscrutable will in creating the desert and in requiring some men to live with it.

Towns like Nishapur, Damghan, Saveh, Qum and Kashan grew up on the routes connecting the principal settlements. Scattered thinly near the foot of the mountains, and where distance on a caravan route across the desert demanded that travellers should find shelter and refreshment, are the villages whose local architecture is the subject of this book (fig 2). The Plateau resembles an archipelago: a sand sea, out of whose barren expanse occasional villages and great ridges of rock protrude like islands (fig 3). All the villagers have built their homes and fashioned their lives in accordance with the same sea. The various settlements knew little of each other until the last decade, which has brought electric light and television to their tea-houses and linked them with the large towns by bus and lorry. They were in fact very similar because climate and geography imposed the same conditions for survival on each of these isolated communities. So it is not unreasonable to infer certain generalisations from the cisterns of Nain, the ice-houses of Abarquh, the wind-catchers and houses of villages near Yazd, the watermills outside Kashan and the windmills of Khaf, even though each is separated by as much as five hundred miles of desert.

In the semi-arid hinterland of these settlements, the raised rim of the sand sea, similar buildings were also erected wherever the amenities they offered suited the climate and the surroundings. So Taft and Sirjan possess ice-houses like those of Yazd and Kerman. Wind-catchers and cisterns are widespread phenomena. Thick mud walls and domed roofs are universal far beyond the desert edge: until higher, cooler climates are reached, where there is readily available stone and timber for a different vernacular style of building. Intermingled in this book will be found buildings from the desert villages and others from its wide rim on the Plateau, for living with the desert has taught men how to live with the dry conditions that prevail all round it.

Because the filaments of the spider's web of Iranian land routes, especially those few that cross the desert, are so fragile and so extended, and because the Plateau lacks the uniting and enriching passage of a Nile, an Indus or a Euphrates, no great urban civilisation ever grew up here. Medieval Isfahan was a fine city, but not in the same league as Baghdad. Magnificent mosques, colleges and bazaars, tiled and domed, were built in Mashhad and Kerman by Timurid princes and wealthy merchants, but the survival of these buildings was almost as precarious as that of the peasant homes elsewhere on the Plateau, alike exposed to destruction by earthquake, drought or raiders from the east.

The desert, pierced by the bone structure of Iran – ridges of serrated mountains whose true height is disguise by the abruptness with which they rise from the desert – dominates and moulds the attitudes of the people. The desert is the real world, while cool courtyard gardens like those of the shrine at Mahan or the pavilion at Bagh-i Fin are so rare as to seem fabulous. The trees and water within the walls that shut out the desert are like a mirage, too good to be true: at any moment the spell may be broken, and brown will reassert itself over this deceptive green. A kill-joy genie will emerge from Aladdin's lamp and send him out of such a transitory paradise back to "the alarming waste without" (Sackville-West 1953, 283). "Cliffs stick up like jagged scorched islands out of the rolling dry desert. Over its dusty surface the dust-devils form pillared halls, and where the heat deceives the eye with mirages the ground bathes in water. Man is an insect, crawling over the roasting hot surface." (Myrdal 1972, 162).

3. Typical desert landscape at the foot of the mountains. Note the spoil mound of a qanat shaft in the foreground and the very sparse vegetation (top). (photo John Donat).
4. The firm gravel surface of the desert near the mountains south of Yazd. Note the defensive corner bastions of the caravanserai, sited beside this important trade route (bottom).

Types of desert on the Plateau

Wherever green has been bleached out of the landscape – which is true of most of the Plateau for much of the year – the western traveller concludes that he is seeing desert. His Iranian companions will confirm his impression by referring to the land as *biaban* (without water) or *dasht* (closely connected with our word 'desert'). There is no other word in Persian for open country, unless man has had a daily hand in its appearance, creating *mazraeh* (farmland , fields). There are, though, several distinct classes of desert on the Plateau, with only the great area of loose sand near its south-eastern edge answering the description of wind-blown, round-shouldered dunes typical of the Sahara and the Empty Quarter of Arabia.

"The conventional view of a desert as a place where rolling sand dunes stretch away interminably to a flat horizon is rudely contradicted in Iran. The landscape south of Firdaus, for example, was composed of a succession of grim and utterly empty mountain ranges, twisted in the most monstrous convolutions, as if the earth in those parts had been smitten with epilepsy and had afterwards relapsed in a set agony. These mountains leapt out of the stony plateau with great suddenness, so that they appeared much higher than the map alleged. There was no break in the drab browns and vermilions of the colour scheme, except for clumps of camel thorn of a funereal yellow green, and occasional dark heaps which were skeletons. Very rarely we met a pilgrim or a string of plodding camels, but for most of the day there was nothing moving to be seen, only quivering heat and desolation inconceivable." (Emanuel 1939, 239).

The term *dasht* should be restricted to the dry, gravelly country, interspersed by short ranges of jagged mountains, which is so characteristic of the Plateau and which can be traversed without undue difficulty by both beast of burden and motor traffic (fig 4). *Kavir*, which has given its name on most maps to the northern part of the desert, is an expanse of viscous mud, frequently overlaid by a salt crust whose varying thickness makes this desert exceedingly treacherous (fig 5). *Namak* (salt) is a salt lake, of which there are several large ones, that may contain open water after winter rainfall. Its white surface flashes a will-o'-the-wisp of a welcome in the east to the traveller bound for the reliable shelter of the caravanserais north and south along his route. *Lut*, sometimes used to describe the entire central desert, but generally restricted to the most barren southern section, means 'bare', 'naked'. It is indeed land without life and it shares this Persian name with another of the harshest features of western Asia, the Dead Sea (*bahr-i lut*). Kavir-i Namak and Dasht-i Lut give Iran its immense dead heart, which occupies c.200,000 square miles or nearly one third of the whole country.

In the twentieth century Sven Hedin and Alfons Gabriel enlarged the topographical and geological knowledge of the remoter tracts of this desert by their intrepid yet unassuming expeditions.

The salinity of the desert has been increasing for millenia due to the action of the sun on the seasonal watercourses. These find no drainage outlet, but are evaporated or sucked down by the parched earth. Therefore the population has been driven to the fringes, where mountain snow provides water for men, for their beasts and crops.

5. The hard salt surface of the Kavir-i Namak.

Crumbling ghost villages remain, spaced out along the few long-established caravan routes across the desert which merchants and pilgrims followed for centuries. In the summer they journeyed mainly by night and sought refuge from the heat of the day in the caravanserais, which offered shade, protection and a vestige of water, sufficient for the occasional caravan but seldom enough for a settlement (fig 6). Marco Polo passed this way. These routes still carry today the sparse traffice from Nain, Yazd or Kerman north towards the great Shia shrine at Mashhad.

A century ago travellers on the main road from Tehran to Mashhad along the northern edge of the desert depended for their survival upon well-appointed caravanserais. At one such, at Miandasht, "Water is abundantly provided. Within the courtyards, and also outside the walls, are some half-dozen large underground tanks, with brick domes built over them to prevent evaporation and keep the water fresh and cool. A flight of forty wide steps gives access to the cisterns, which are closely guarded by the caretakers of the establishment. It is true each traveller is allowed all he needs for his personal use, but the fluid cannot be wasted, nor is it allowed to be given to the animals" (O'Donovan 1882, I, 421). There were tanks of poorer quality for the latter some way off, but the subterranean channels feeding all these cisterns dried up in the summer. Only careful forethought and discipline enable men to live with the desert.

A climate of extremes

The minerals in the rocks and sands of the Plateau produce marvellous colour early and late in the day, but – with the infrequent exceptions noted below – there is no trace of moisture in the landscape. It has been desiccated beyond redemption by the sun and wind. The medieval Arab geographers divided Iran into the *garmsir*, the Hot Lands, and the *sardsir*, the Cold Lands, by an imaginary line drawn east to west, roughly following the southern boundary of the Plateau from Baluchistan to the head of the Persian Gulf. The bitterly cold winter, turning to snow in the mountains and occasionally in the desert too, justifies the Plateau's inclusion in the *sardsir*. However, its summer temperatures almost match those of the Persian Gulf and in places surpass them. In midwinter the temperature in the central desert can range from -5°C to +40°C (23° - 104°F) in the space of the six morning hours (Redard 1966, 5). Man's survival has depended in part on his ability to protect himself from these extremes of heat and cold and to turn them to his advantage where possible. In South Australia the similarly intense heat has led to the construction of completely underground houses for the opal miners, as at Coober Pedy and Andamooka. Other pre-scientific architectural tricks have made life more comfortable on the Iranian Plateau, as later chapters in this book will demonstrate.

The aridity of the land is maintained and emphasized by the wind whose prevailing direction over much of the Plateau is from the north-east. The notorious '120-day wind' blows through eastern Khorasan and Sistan during the summer months, gusting regularly at Gale Force 9

6. A caravanserai near Semnan on the northern edge of the Kavir-i Namak. Photograph taken from the ruins of an earlier caravanserai to emphasize the Persian habit of deliberately not rebuilding on the same site. (see P. 16).

and sometimes reaching Hurricane Force 12. Small wonder that it is also nicknamed 'the black wind' and 'the wind that kills oxen'. Its greatest intensity coincides with the heat and somnolence of the afternoon, when man retires indoors and sleeps it off as best he may. Outside the gale scours with grit all that stands in its path. The wind-eroded rock formations of the Shahr-i Lut in the southern desert are echoed by the ghost cities of medieval Sistan wrecked by Tamerlane in 1383 and ever since mouldering to dust under the lash of the wind. One province of north-west Afghanistan, Badghis, and one region of Khorasan, Bakharz, are specifically named after the wind (*bad*). Settlements are orientated for shelter from the wind, whilst also seeking to tap its power in certain ways. At Yazd the winds are known by distinct names and carefully borne in mind by builders: *shomali* from the north, *khorasani* from the north-east, *kermani* from the south-east and *isfahani* from the south-west. The lie of the land around any village naturally affects the precise direction and strength of the wind that besieges it in different seasons. The 120-day wind, blowing from the north-east in Khorasan, has veered to the north-west by the time it reaches Sistan, due to the high ground around Birjand and Qain, and then peters out across the border in Baluchistan.

Water: meagre resources at man's disposal

The parching quality of the wind, combined with the summer heat and the scanty rainfall on the Plateau, makes the conservation of water vital. The annual precipitation over much of the Plateau seldom exceeds 126 mm. (5ins) and may well fall in one or two torrential storms when houses collapse and the meagre top-soil is washed away. Irrigation must be practised for any reliable harvest, and for this reason the Iranian peasant has acquired a remarkable expertise in the channelling and distribution of water, as well as a tenacity in the defence of his water rights (see Chapter III, P. 39).

No rivers reach the sea from the Plateau and only three really sizeable rivers flow across corners of it for any distance: the Zayandeh Rud which creates and sustains Isfahan before seeping away in a swamp eighty miles further east, the Hari Rud which serves Herat and is doomed to trickle away in the sands of the Kara Kum desert, and the Helmand which swells the Hamun in Sistan in the spring. Water drawn from these rivers ensures the agricultural prosperity of the long oases upstream from Isfahan and downstream from Herat. No more than five miles from each river, however, the greenery gives way abruptly to the desert: there is a limit to the ability of irrigation ditches to carry the river water into barren, unexploited territory. Sistan, in an era of more extensive agricultural settlement, was once termed "the granary of Persia." The spring floodwater in the arms of the Helmand delta was channelled over a wide area and enough grain was grown to supply west Afghanistan and eastern Iran as far as Yazd, a 24-day journey for camels laden with wheat (Napier 1876, 194).

This irrigated grain, grown wherever possible, is the peasant's main concern. Away from these few rivers, which provide enough water for rice cultivation as well, fields are few and the green or gold strip between oasis village and desert is always slender. The sudden appearance of agricultural prosperity in the midst of a desert landscape is explained by the straight lines of giant mole-hills or bomb-craters heading from the distant mountains. This is the *qqnat* system (see Chapter III) that creates the oasis culture which is a key feature of the Plateau away from the immediate proximity of mountains. Every drop of water counts and is preserved by shading it from the sun in tunnels, cisterns and in the shelter of trees until it is spread through a network of miniature channels over the fields it has helped to create. Khaf, a small town in Khorasan typically dependent upon qanats, could support about 4,500 families a century ago, but had no surplus grain to export. In any case two thirds of the population needed to live as semi-nomads in the surrounding desert, searching out the scrubby vegetation to feed their sheep and goats (Napier 1876, 201). Deep wells have to some extend transformed the agricultural face of parts of the Plateau in recent years, but the long-term success of such methods is very uncertain. By contrast, disputes with Afghanistan over the Helmand river have caused much of Sistan to revert to desert.

Trees against the desert

Vegetation that holds man and the soil together, keeping the desert at bay, is thus largely due to irrigation. What little natural growth there is, dependent upon occasional rainfall, occurs mainly round the Plateau rim, where grazing animals make short work of any vegetation. There is no thick ground-cover, nor rank weeds to compete with the corn. "Every blade in the artificial wheat field is an isolated unit " (Malcolm 1905, 4). The peasants are conscientious husbandmen of their water-supply, planting saplings beside the irrigation channels and growing both lucerne and fruit trees in those small mud-walled fields which extend as far as the water allows beyond the central huddle of houses, pushing back the desert. These are the true Persian gardens, rather than the exceptional, formal, aristocratic creations surviving from the seventeenth century at Isfahan and Fin and the nineteenth century at Shiraz and Tehran.

Most trees planted alongside water-courses are poplars, but in places on the Plateau various other trees surprise the traveller: pines in the neighbourhood of Khaf, the great planes by the shrine at Natanz, the date groves of Bam and Tabas Golshan, the occasional tall tamarisks – the only wild trees to be found in the central desert – of the south-eastern Lut. Pistachio trees, common and

7. Only a mud wall separates this well-planted garden near Mahan from the desert. Poplars are grown for roofing timbers. The domes of the village bath-house are surmounted by ventilators (see P. 26). (photo Robert Harding).

growing wild just beyond the eastern edge of the Plateau in north-west Afghanistan and extensively cultivated around Rafsinjan, near Kerman, were probably once much more widespread on the Plateau. Grazing animals, temperamental rainfall and man's search for firewood have aided the sun and the wind in denuding the scanty cover of large areas of the Plateau which might otherwise support very limited agriculture independent of irrigation.

The shade, the coolness and the greenery of orchards – a more accurate word than 'garden' for the kind of Persian *bagh* found round all small towns and villages on the Plateau – have two vital contributions to make to success in living with the desert. Firstly, they rejoice the owner's heart: his family and friends find refreshment there every Friday as they lounge in pyjamas on rugs under the trees with picnic bundles, samovars and playing cards. Secondly, these gardens help to lay some of the dust which the wind hurls at the settlements standing in its path. Where the wind wins the day, perhaps because man neglects to repair the sheltering walls or because the water-supply fails, the garden reverts to waste, rendering life nearer the village centre that little bit more intolerable. Medieval Arab geographers refer to one novel use of wind-towers in Sistan: to swirl up encroaching sand and deposit it at some distance (Forbes 1955, 113). Surely a rather desperate and unreliable last ditch measure if the desert was already overwhelming one's orchard?

Enemies of agriculture

The abrupt transition from cultivation to desert, a marked feature of the Plateau landscape particularly attractive to photographers, is a continual reminder of the harshness of human existence on the Plateau's terms (fig 7). Agriculture has remained unchanged for many centuries partly because the peasants knew they could not ask for more than subsistence from their soil. Each plot had to be tended and watered with continuous care, otherwise the sun and the wind would quickly reimpose their parching sway over it. Since one third of the harvest was often taken by the landowner, who levied similar oppressive taxes on the animals and the paltry possessions of the peasants, there was little incentive to raise production.

In medieval times the taxes of the Mongols bore heavily on a peasantry whom they had already terrified, decimated and ruined, for they required the survivors to pay twice as often as the nomadic tribes. Later the standing army of the Qajar shahs and the armed retainers of the great landowners, who took what they wanted and against whose depredations the peasants had no redress, were more frequently to be feared than a plague of real locusts. The peasants were thus forced to the verge of famine, to that dread decision to eat the seed corn and condemn themselves to an uncertain future, mortgaged even further to their landowner. Scant snowfall in the winter, causing a decreased flow in the qanats next season, might then spell starvation. Outbreaks of diseases like cholera, typhus and smallpox and periodical, calamitous earthquakes seemed to indicate that even God's wrath deliberately sought out the weak of the earth. So nobody outside the extended family circle was to be trusted and one counted one's luck from day to day.

Prosperity was felt to be the lot of the merchants; the peasants could only cherish the short-term ambition of surviving another winter. The sole exceptions were probably those cultivating the Zayandeh Rud and Hari Rud valleys, where nature was prodigal with water, so that fruit as well as grain could be grown for the markets of Isfahan and Herat. Those who clung to life closer to the desert kept their heads down, worked hard and hoped for the best: a bellyful of bread was the summit of most aspirations in the Plateau villages.

Measures of self-defense

Another reason for the sparseness of settlements on the eastern edge of the Plateau is the nineteenth-century Turkmen raids, which ravaged villages more slowly and systematically but just as effectively as those of their Mongol predecessors. The supine attitude of the peasantry and of the Qajar government enabled these horsemen to sweep down from Khiva on razzias for several hundred miles across the Persian frontier, carrying off animals and prisoners to the slave markets of Central Asia. They followed the route offered by geography: between the eastern extension of the Alborz chain and the deserts of western Afghanistan. Since raiders depended upon surprise they travelled too fast to transport artillery, so the film set of crenellated mud brick towers and walls of Bam (fig 8) was able to withstand several Afghan attacks in the 1840s, while the fort at Neh was still a place of refuge from Turkmen, Afghans and Baluchis within living memory.

On the fertile, well-cultivated northern edge of the Plateau, where the Turkmen were much feared, fortified villages stood no more than half a mile from each other, with strong walls, "in some cases triple, the outermost being moreover protected by a deep ditch or a wet moat. All the walls are furnished with flanking towers, and, together with the central buildings and gateways, are elaborately ornamented with mouldings, made in the plastic clay of which the whole house is built. In fact, the ornamentation is stamped on somewhat in the same way that a pastrycook manipulates pie-crust, and often reminds one of the latter in its appearance" (O'Donovan 1882, I, 383).

The stretch of road between Shahrud and Mashhad was known as 'the Marches of Terror' and as late as 1928 the Turkmen threat was so real that the Persian Government retaliated for their marauding incursions by despatching aircraft to bomb their camps along the north-east frontier. Travellers were wise to travel in caravans and to employ a posse of armed guards (Filmer 1937, 161).

Peasants working in the fields stood little chance in the event of a well-executed raid, unless the village operated a well-organised watch system. The threatened people might then scurry to safety within their small citadel, rolling into place the great millstone that still exists in villages near the Afghan frontier like Niazabad. Unlike wooden gates this could not be set alight or easily battered down by determined bandits.

8. Barracks and stables within the deserted citadel and old walled town of Bam (1977).

Another feature of this north-eastern area of Khorasan, recorded on the Survey of India maps of sixty years ago, were mud brick refuge towers which have long since crumbled into the fields they protected. They were vividly described in an 1881 lecture to the Royal Geographical Society by a British Army intelligence officer. Near the border "the refuge towers are very close together, every square of 150 yards of the fields having one. The whole

9. Pigeon-towers near Shahreza in the 17th century (Chardin's "Voyages", III, Paris 1811).

10. Bastion and town walls of Yazd, dating from the 14th century and kept in repair until the 20th century. (1977).

country here is so thickly dotted with them as to look like a chessboard covered with chessmen. The towers are small, round buildings, built of unbaked clay, about 12 feet high. They are roofed over, and have no opening whatever except a small round hole at the bottom, through which a not-too-stout person may wriggle himself in like a snake. If surprised by Turkomans, the cultivator or traveller creeps through this hole, and closes it with two large stones, which are there for the purpose. Even if these stones are wanting, the occupant is safe, as it would indeed be a daring Turkoman who would try and force himself through the hole, with the certainty of having his brains beaten out with a stone while struggling to get through, even supposing the person inside had no better weapon, but almost everyone here goes armed. The defence towers are higher and larger, and have a parapet at the top, with loopholes to fire through, and a ladder for ascending to the top. Each vineyard or orchard has its one or more towers" (Stewart 1911, 397).

One wonders whether the pigeon towers in the neighbourhood of Isfahan were turned to similar use in uncertain times, especially during the great Afghan invasion of 1722 (Fig. 9). Mud towers built for the summer watchmen and bird-scarers in fields and vineyards afforded makeshift observation platform and temporary sanctuary under similar circumstances, as also did the wind-mills on the northern side of certain villages in Khorasan.

However, the Persian peasants of a century ago went

11. A wayside shrine north-west of Kashan. The barren plain continues right up to the walled enclosure, where pilgrims to Qom can pause for shade and water. (photo John Donat)

in abject terror of the Turkmen raiders and seldom showed much martial spirit. There was no lasting safety except behind impressive walls like those of Yazd (fig. 10), where the provincial governor would maintain sufficient troops to discourage raiders and brigands. The drift to the towns, which one is tempted to reckon a modern phenomenon in Iran, has been an important feature of life on the eastern side of the Plateau for several centuries.

*   *   *

The reverse side of this terrible coin of oppression and misery was a resourceful attitude towards those daily elements, the sun and the wind, whose occurrence and effects could be confidently predicted.

The buildings that resulted from this attitude, and that are described in later chapters, do not compete with the desolation of the desert landscape. The few trees and fields that surround most settlements make that challenge. The buildings belong to the landscape, for their varied domes and towers that break the skyline in each village echo on a human scale and in gentle and regular geometry some of the colours and shapes of the hills behind them, even of the saw-toothed mountains that so often close the horizon on the Plateau (figs.11, 12).

Man, doomed at times in summer to crawl insect-like over the roasting surface of this inhospitable region, finds shelter, hope and dignity as he enters the shade of his own threshold.

12. Roofs in Khurunaq echo the shapes of the desert hills beyond the dry river bed. The domes, chimneys and glass-framed vents in the centre indicate the village hammam *below. The fields are carefully laid out to take best advantage of any qanat water available for irrigation, some of it carried in the aqueduct visible here.*

2 BUILDINGS

Most of the buildings considered in this book could be described as vernacular, the term being used as it is in connection with language: vernacular language is that spoken by the people; vernacular building is that built by the people. Architecture without drawings might be another way of describing much of it.

Although such buildings may have few architectural pretensions, they often possess architectural qualities of a very high order. Sir Henry Wotton's famous summary "*Well-building* hath three conditions: *Commodity*, *Firmnesse* and *Delight*" is a sure yardstick against which to measure this claim. (Wotton 1651, 201).

Commodity, the usefulness and convenience of a building which result from its design for a particular purpose, is discussed in Chapters IV – VIII. In this chapter that other aspect of *Commodity*, the provision of shelter from the elements, together with *Firmnesse*, the ability to remain standing, are discussed. Most buildings, sharing the same climate and terrain and using local materials, solve these problems in the same way whatever their precise function.

Delight, Wotton's third Condition, comes as a direct result of the fulfilment of his first two in the context of the vernacular. It is none the less genuine for this. Indeed, delight is the emotion which first makes the stranger stop and stare. Many of these buildings are astonishingly beautiful; others are wonderfully curious; some are both. Those on the Iranian Plateau are sometimes monumental both in scale and size (see, for instance, Chapter IV, Ice-houses); this comes as a surprise after the more typically humble vernacular buildings elsewhere.

13. Flat roofs in a Khorasan village. Wherever there is enough timber for their construction most buildings, except mosques and shrines, have flat roofs.

Mud

Vernacular buildings always rely on the most readily available material and this means mud over nearly the whole of the Plateau where it is used almost universally for walls in one form or another. This is also true of much of the Middle East and of huge areas of Africa, Central America and India, where, given a dry climate, strong similarities exist in their buildings. In wet parts of Iran (e.g. on the shores of the Caspian and on the northern slopes of the Alborz Mountains) there is plenty of timber, and buildings are designed to shed the rains. Structures with deep-eaved sloping roofs are typical, as they are in sub-tropical conditions elsewhere.

Roofs of the Plateau buildings, however, are of two kinds. Where timber is available they are flat, with timber joists supporting thick layers of mud laid on boarding (fig 14). Elsewhere mud brick or baked brick have been used structurally to roof buildings of every kind. Myriads of domes and vaults of astonishing sculptural quality spring up over humble village houses, over work-a-day structures like cisterns or mills - even over animal shelters (fig 15). They are not reserved for shrines or mosques as they might be for churches or cathedrals in Europe.

Bearing in mind that words mean infinitely more to the Iranian than mere fact, it is peculiarly satisfactory that the word for rose and that for mud are written (although not pronounced) in the same way in Persian. Nor was the universal importance of mud forgotten by the great poet, Firdausi, who in his national epic, the *Shahnahmeh* (The Book of Kings), describes how the hero Jamshid (son of the first man) "taught the unholy demon tribe to mingle water and clay, with which,

14. A flat-roofed grape-drying shed in Khorasan. The poplar plantation which provided the poles from which the roof was constructed can be seen in the background. The ventilation holes make an effective pattern on the rear wall. (Similar but much grander grape-drying sheds are built to 2nd storey level in villages north of Kabul).

formed into bricks, the walls were built, and then high turrets, towers and balconies, and roofs, to keep out rain and cold, and sunshine" (quoted by Simpson 1892, 698). Poetic licence may have inspired Firdausi to push the invention of mud brick further back than the archaeologist might approve but it has been highly regarded as a building material for several thousand years. The first walls of Jericho were built of mud. And it has been proved to be a suitably durable basis for the earliest paintings so far discovered on man-made (as against cave) walls. Those in the shrines at Çatal Hüyük in Anatolia are on plastered mud brick and date from the seventh millenium.

Given protection, mud is not only durable, it has qualities which make it oustandingly suited to the harsh desert climate of extreme heat, cold and glare. Only in the mid-twentieth century has the vogue for the worst the west can offer, bad imitations of bad imitations of modern architecture, penetrated much of Iran, ousting the vernacular. It is often forgotten that the first modern buildings, Le Corbusier's early villas (1929), were designed for a temperate climate in an energy-rich society – conditions diametrically opposed to those now enjoyed on the Plateau. Yet to build in the undigested "modern" manner bestows prestige. This trend is common enough in English villages where the thin wall and big window, though often a visual disaster, may be tolerable in terms of day-to-day living for at least part of the year. But in extreme climates they are not, and they can only be made so at great expense. Yet mud, despite its aesthetic and practical advantages, is still largely out of fashion. It is in all of us to want to try something new, forgetting the virtues of the old and remembering only its weaknesses. Inevitably mud building in most of the third world is

15. Vaulted roofs in a village near Isfahan. (photo John Donat).

inextricably associated with poverty and overcrowding, so those, including westerners, who have enjoyed living in comfortable mud brick houses in Middle Eastern villages find it difficult to persuade others of their virtues, virtues which must be appreciated if mud building is to be considered.

A slum is created by the way a building has to be lived in as much as the building itself. Some of the worst slums of London and Liverpool were once upper-class Georgian terraces while countless artisan dwellings are now being rescued by the middle-classes. It is important to think of a building material objectively and not to be confused by its social context.

It is equally important to consider how a material is used, rather than the material itself in isolation. Mud is probably more dependent than most on the skill of the builder. Its plasticity demands a remarkable degree of control if a building of which it is constructed is to remain standing. Mud pies are a delight to make but they are not durable. However, many a Plateau villager can build as well as he can farm or ply his trade and he is not alone in his appreciation of the potential of the material. Surprisingly perhaps, British forces in East Persia learnt to build in mud in the 1920s (Dickson 1924, 142) and recently in South Australia, the practice of building in mud has been enjoying a revival (see bibliography). Weekend courses on the making of mud bricks and on construction in earth and *pisé* has had the practical result of giving people a chance to build for themselves. It is worth noting that the thermal properties of such buildings have been among their chief attractions.

Thermal insulation

Land at an altitude of 1,000 - 2,000 metres (1,000-6,000 feet) flanked by high mountains usually has a desert climate with extremes of heat and cold, ranging from below freezing to over 38° C, (c.100° F). It is against such extremes of temperature that protection is most needed. It happens that mud is an excellent insulator provided it is dry. It is a relatively porous material so the minute air spaces within it make it a much better insulator than stronger, denser materials like baked brick or stone. The mud wall retains the heat instead of allowing it to pass quickly through.

The thermal conductivity of a material, known as its K value, is a measure of heat transference through it (in terms of unit time, thickness, area and temperature difference); the lower the K value, the better the material is as an insulator. For instance:

Insulating fibre-boards have a K value of around	0.053
Mud brick	0.22 - 0.32
Stabilized soil blocks	0.5 - 0.7
Baked brick	0.71 - 1.45
Concrete (mix 1:2:4)	1.44
Limestone	1.53

These figures show mud brick to be a very much better insulator than, for example, concrete. Furthermore mud walls have the additional advantage in this respect of being thick. They could not remain standing otherwise but would lean, bend and collapse since mud and mud brick have no tensile strength. This generous thickness is not an extravagance since construction is cheap and the material in abundant supply. And so a material which happens to have a low K value is used lavishly thus giving exceptionally good thermal insulation.

Conversely, walls built of concrete tend to be thin. This is structurally possible since the tensile stresses typical of any wall are taken by steel reinforcement either in the thickness of the wall itself, or more often, by a framework of reinforced concrete or steel columns and beams within which the wall is simply a thin panel infil – an inevitably expensive method of construction. The additional thermal insulation (e.g. fibreglass) which is essential for comfort may be skimped.

The huge thickness of the normal mud wall or roof has another important advantage. This material has an exceptional capacity for heat storage: the thicker the wall or roof the greater its storage capacity. In any desert climate, very hot in the day and very cold at night, the clear sky allows the earth (or the building on it) to receive a large amount of radiant heat from the sun which is radiated back into the cold, starry sky at night. Some of this heat is also radiated back into the rooms, the whole house being, in effect, a storage heater. For much of the year, when any heat, particularly during the hours of darkness, is an asset, this is valuable. The temperature of indoor wall surfaces makes a great difference to comfort since when these surfaces radiate heat to the occupants of a room, they feel comfortable even though the actual air temperature of that room is below comfort level. (10% - 15% less radiant heat than convected heat is said to be needed for comfort). During summer nights when the rooms of a mud brick house are inevitably too hot, it is the custom to sleep on the flat roof – a fine place to be once the glare of the sun has gone.

Maintenance

No material is without its drawbacks and mud has some major disadvantages: its resistance to rain, snow and earthquakes is poor.

The traditional Iranian attitude to rain and snow is eminently practical. Both are comparatively rare and can be countered by instant maintenance rather than anticipated by prohibitively expensive design, so roofs are made to be easily accessible. The typical western approach on the other hand, transplanted to Iran from a temperate climate, tends to give water-proofing priority over the constant need for efficient thermal insulation – to the great discomfort of those using the building.

The Persian's practical attitude makes even more sense when it is remembered that a scant "average annual

16. Maintenance steps make the dome of a desert shelter easily accessible.

rainfall" of, for example, 125 mm (c 5 inches), may not fall at all in some years but be concentrated into half a dozen falls in others. The occasional disaster does of course occur. Napier Malcolm who spent five years as a missionary in Yazd around 1900, described twenty-four hours' rain which destroyed a couple of hundred roofs (Malcolm 1905, 29). Who could afford to build to counter such rare extremes? During very wet weather on the excavations at Pasargadae in the autumn of 1963 those workmen who came from distant villages asked to go home to see if their houses were still standing: this was simply an act of common prudence. Happily, none had collapsed but the houses could have deteriorated quickly without attention.

Snow too has devastating effects if it thaws on mud roofs, so wide-bladed wooden shovels are kept at the ready to hurl it off. A walk down a village street has its hazards after a snow-storm and this is as true in big cities like Isfahan. Stone rollers also are kept on roofs to compact the mud whenever it has been softened by snow.

Most flat roofs are easily reached by steps either from indoors or from outside because of their use for sleeping on summer nights, or for fruit drying and storage, inaccessible to animals. Access to domes is often by ziggurat-like ramps which spiral up them (fig 16), or by intermittent steps. At first sight these look like applied decoration but they are primarily designed for maintenance.

However, despite such practical evidence of the recognition of the need for instant maintenance, it is often left undone. A seventeenth century traveller, the Frenchman Thévenot, was mildly exasperated by the Persian's lackadaisical approach: "They take no care to keep things in repair, and it happens many times that they will let a whole building run to ruin for want of a handful of Plaster" (Lovell 1687, 18). A stitch in time... But after tempest or flood most buildings need maintenance, and they need it quickly since at this stage most are potential ruins. Speed is vital and this is not possible when so many buildings need attention simultaneously, as they must. It may be relatively simple to attend to the roof over your own head or the mill which you are actually using, but a building that is even temporarily unoccupied must have lower priority. Indeed, it may be cut off by a raging river or a snowdrift. Deterioration is then fast and inevitable.

Destruction by earthquakes is another matter. In that which flattened Tabas on 15th September 1978, thirteen thousand of the city's fifteen thousand inhabitants are reported to have been buried alive as they were having supper. Forty of the surrounding desert villages were also destroyed.

17. Ruined buildings beside new. Abandoned buildings are a common sight; it is usually simpler and more economical to build from scratch than to repair. Anarak, 1975.

Having no tensile strength mud buildings are not expected to survive earthquakes. From the survivors' point of view, village houses at least have the virtue of being cheap and replaceable. Timber has been introduced in some buildings to give tensile strength, often rather half-heartedly; but the general feeling seems to be that little can be done on such occasions to save a building or its occupants. An earthquake is a fact of life which is in the hands of Allah. (Research into the performance of mud brick during earthquakes is going on, although nothing as yet seems to have been published. One such project is that on the use of stabilised soil building products which is being conducted by the Overseas Division of the Building Research Establishment within the Department of the Environment.).

Following an earthquake re-building is obviously essential but when mud buildings have not been repaired quickly after rain or snow it is also usually cheaper and safer to build new rather than to repair. Traditionally there has been an abundance of cheap raw materials and a large labour force skilled in using them. The availability of building land is another factor. A relatively small population living on the fringe of vast tracts of virtual desert has plenty of scope in this respect. So much space means that ruins are seldom demolished to clear the site and new buildings are put up nearby.

Such practical considerations may have combined to produce a tradition in which the idea of preserving a building, unless it is sacred, is seldom considered; an attitude which tends to surprise westerners. But the interest in the conservation of historic buildings only developed in Britain comparatively recently. The relish with which the seventeenth and eighteenth century English gentleman updated his ancestral home tends to be forgotten. How many trim Georgian facades mask an accretion of earlier building?

Certainly, Sir John Chardin, the court jeweller who was on business in Isfahan in the late seventeenth century, was sympathetic to the Persian attitude: ".... The Persians do not like their Parents' Houses, they love to build some fit for themselves, which is very rational: for, as they say, there is the same difference, between building a House fit for one's Family, or taking one ready built, as between making oneself a Suit of Cloaths, or buying one ready made." (Chardin 1927, 264). Malcolm noted that "to the Yezdi an old house means a bad house", and he pointed out that a new house was not intended to outlive the life of its cess-pit – forty or fifty years. He went on to give the opinion that two thirds of a Persian town or village was ruinous (Malcolm 1905, 27, 28).

Less fundamental than such practical aspects is that of changing architectural style. Fashions of style never have much effect on vernacular building and where a decrepit building is to be replaced by one which is similar in both style and materials, the architectural reasons for its conservation have often largely gone. Persian builders have seldom favoured innovation for its own sake and their time scale is very long (making buildings without inscriptions difficult to date).

An additional and positive reason for this lack of interest in old buildings is the pleasure and reputation to be derived from putting up new, whatever the style. Foreigners could be tart about this, seemingly forgetting what went on at home. John Fryer, a surgeon of the East India Company, who was in Persia about the same time as Chardin, described new bridges as being built "chiefly for Ostentation, or to Preserve their (the Patron's) Memories, as other great Buildings are". Without such vanity the world would be a deal duller and unemployment worse.

John Fryer also noted a great virture of the mud house ".... they have a better opportunity against Fire their Mansions affording little combustible Substance for that to feed on". (Fryer III 1912, 17). This too may go some way to contributing to the surprising numbers of ruins.

Whatever the reason behind a new structure, the building of it gave the Persians great satisfaction. The urge to build in a country largely devoid of good stone or timber is surprisingly strong. Perhaps the temporary nature of so much shelter in the Islamic world, be it a mud-brick house or a goat-hair tent, has given special importance to the builder of more permanent structures. A sign of the coming of the Day of Judgement foretold by the Angel Gabriel (related in the Second Hadith) runs "You will see the barefooted, destitute, naked herdsmen competing in constructing lofty buildings".

An integration of the spiritual with the workaday may go some way to explaining the apparent extravagance of building patronage which has repeatedly amazed European travellers, finding as they do, comparatively new buildings in various stages of ruin, superseded by a new structure of nearly identical design.

Plateau soils

At an altitude of 1,000 metres or more the soil is typically sandy earth with clay. It is often greyish, being calcareous throughout with an accumulation of soluble salts, or it may be a crumbly brown soil which grades into a pale brown or greyish, highly calcareous clay 15 - 30 cms (c. 10 to 20 inches) below the surface (Cambridge History of Iran 1968, 256). There is very little organic matter in either soil and this, however unconducive to agriculture, is an obvious advantage to those who wish to build. The earth from which a building is constructed is usually dug on the spot, as it once was in England, but on the Iranian Plateau there is seldom top soil to be moved. Alternatively if the new buildings is alongside a ruin, the old earth walls may be used as a "quarry".

The earth differs from place to place in its proportions of clay to sand. This proportion decides how it is best used, knowledge which is traditional to the village builder. Rammed earth (*pisé*) requires over 50% of sand, a method which is rarely used on the Plateau. Mud walls and mudbrick (which are by far the most common) are best made with over 50% of clay. Earth for baked bricks usually has 70 - 80% clay.

Mud walls *(gel)*

For mud walls the earth is used wet, as mud, like the cob once common in Britain: the earth is very thoroughly mixed with the water and with short bits of straw by being trodden by bare feet - a pleasant job in hot, dry weather. Fortunately the straw left by the threshing sledges is mostly broken into pieces approaching the right size. Its function is to equalise the shrinkage as the mud dries out. Straw is at its cheapest after harvest so building may wait until then. At all events it must wait until there is not much to be done on the land so that neighbours can help each other over any building work that is needed.

The Persian mud wall and the English cob wall seem to have been built in much the same way. One man with a long handled shovel throws the mud up to another who spreads it on the wall (fig 18). Or the mud may be made into balls which are tossed up and packed together on the wall. Occasionally the mud balls are contained in timber formwork, a method seen in Khorasan, but this seems to be uncommon. Construction varies according to local conditions but skill in handling the material is universal: "The cleverness of the Yazdi in manipulating mud is beyond belief .... Many even eat mud, and develop unwholesome muddy complexions." (Malcom 1905, 12).

The height of the section of wall, or lift, which can be built at one time is about 50 - 70 cms (c. 20 - 30 in). After this it must be left to dry out for a couple of days (depending on the wetness of the mix and the dryness of the air), otherwise the wall will collapse under its own weight. Each lift is usually battered, giving a slight drip to throw off water at the lift line. The horizontal lift lines can easily be seen and are similar to those on English walls.

These walls differ from those made of earth in France, known as *pisé*. *Pisé* has no water added. The earth is piled between wooden shuttering and then ram-

18. *Building a mud wall. The earth from which this wall is being constructed is dug alongside. The lift line can be clearly seen below the shovel.*

med. Nor is straw needed because there will be no shrinkage to counteract. The "dry" earth from which *pisé* walls are built is likely to be a good deal damper in France than that which might be dug in the desert where the exceptionally dry atmosphere would dry the earth very quickly once it was exposed and make it unworkable whatever its clay content. A disadvantage of pisé in treeless country is the expense of timber shuttering.

Mud walls 10 metres (33 feet) high were commonly built as shade walls for ice-making (fig 57) or for the protection of villages or rich gardens against advancing sand as much as from human trespass. These walls are buttressed as needed and garden walls often have small turrets which have the additional function of buttressing (fig 20). Such walls would be at least 2 metres (c. 6 feet) thick at the bottom and are battered. A wall of 2½ metres (c. 8 feet) height might be about 50 cms (c. 24 ins) thick at the base.

Like the Devon cob wall, Persian mud walls need "a good hat and a good pair of shoes" because once water gets in, disintegration follows quickly. Disintegration is accelerated by the freezing winters common in the desert: the expansion of any water which may have got into the wall soon destroys it.

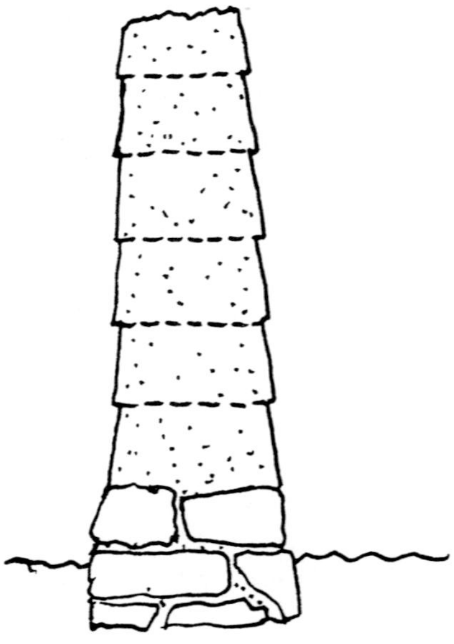

19 Section through a mud wall showing its lift lines

20. A mud wall of a garden near Abarquh. The lift lines here have been lined with bricks. Note the brickwork pattern on the corner bastion.

Walls are usually protected at the foot by being built on a rough stone foundation which is bound by a lime and mud mortar. The "hat" or coping may simply be a continuation over the wall's top of the plaster (see below) with which the face of the wall is coated. In some areas a course of projecting burnt bricks forms a drip to throw the water clear. Ordinary garden walls are often topped by a row of projecting sticks on which a mass of desert thorn is held down by more mud mortar. This thorn has the additional advantage of deterring trespassers.

The time that must elapse between lifts, prolongs the building of a wall. This may account for the introduction of sun-dried bricks (see below), sometimes it seems, at random, but at others to a deliberate scheme. Jean Baptiste Tavernier, a French diamond merchant and most observant traveller, who was in Isfahan in the seventeenth century, showed great interest in "the manner of building in Persia" and noted that "the Walls are made of lays of Earth, according to the proportion intended, and between every lay of Earth, three Foot high, two or three rows of Bricks bak'd in the Sun". (Tavernier 1678, 147). The upper part of a high earth wall may be built of brick to overcome the difficulty of getting the mud so high, or because the top part of the wall needs less strength so need not be as thick as cob construction would dictate.

Mud bricks *(khesht)*

Mud bricks are made of a similar mix of earth, water and short straw as that which is used for mud walls, but slightly softer. The mud is rapidly shaped in a wooden former (see fig. 21 on hard flat ground on which it is knocked out to be dried in the sun. After initial drying on the spot where they are cast, the bricks are closely spaced on edge in neat rows, or sometimes in sinuous lines, to dry for two or three days (see fig. 21). The object is to get as many as is convenient into the space. They are more like thick tiles than western bricks 20 x 20 x 4 or 5 cms (8 x 8 x 1½ or 2 in) is typical; oblong are also made. Tavernier describes similar bricks which were "three fingers thick" (Tavernier 1678, 147). This size is easy to handle and the brick's large surface area (roughly double that of a western brick) provides a generous surface area for mud mortar. This is important in vault construction (see below) during which there are times when the bricklayer relies on the sticking strength of the joint. The thick mortar joints are of the same mud mix as the brick. This is an ideal situation which English bricklayers using fired bricks can but envy because the brick and its joint will weather evenly. When a wall is out of level (perhaps because of the unequal size of the rough foundation stones, or the slope of the ground), bricks may be laid diagonally to bring the wall to a horizontal course.

21. Making mud bricks (top) Roughly western shape. (middle) Tile shape. (bottom) Drying bricks in a serpentine line.

Lengths of timber are sometimes built into mud brick walls to take the tensile stresses in case of earthquakes (it would be logical to construct half-timbered buildings, using the mud brick as infill only, but timber is far too scarce). Similar timbers may also occur in domes, presumably for the same reason. The ice-house in fig. 57 (P. 54) has a ring of timbers, 30cms (12 ins) in diameter built into its dome about 5 metres (c. 16 feet) above ground level.

The compressive strength of mud brick will vary according to its clay/sand content but is relatively unimportant since the complete lack of tensile strength both in the brick and the mortar joint necessitates the building of thick walls. A concrete block would have two to three times the compressive strength of mud brick.

Baked brick ***(ajur)***

The use of baked bricks was once confined to buildings of some architectural pretensions. It is obviously more durable and it also gives status and it is now often used for workaday structures and ordinary village houses. Walls are usually left unplastered outside since there is no practical need to plaster them but the brickwork tends to look harsh. Domes and vaults, on the other hand, are usually plastered outside but often not inside and fascinating patterns are produced in the brick bond.

Gypsum *(gach)* mortar is generally used with baked brick. For vaulting this has the advantage of being very quick setting (see P. 22).

Baked brick is also used in an entirely practical way where a building may be vulnerable to damp or subject to knocks or wear and tear. Then little inhibition is shown about the *ad hoc* introduction of baked brick into a mud brick wall wherever it seems convenient. The builder will use what is to hand. Sometimes baked bricks are used instead of stone for foundations. Fryer noted this when describing how snow, thrown into the street, thawed and undermined the walls of houses which " .... the Rich better secure by Brick Bottoms, than which the Poor, whose Fabricks often totter on such occasions". (Fryer III 1912, 16).

Brick is used for paving as well as for walls. Being relatively cheap, it is used more often than stone to pave the courtyards of houses. Roofs may be paved too for protection against the wear of people's feet as much as for weather-proofness.

Baked bricks are often made from the same mould as sun-dried bricks but since they shrink in firing they end up smaller. Oblong bricks of rather more western proportion, 20 x 10 x 5 cms (c. 8 x 4 x 2 ins), are also made. Small kilns, now fired by oil, still produce bricks for village work (fig 23).

22. A brick dome (in a caravanserai at Khurunaq, 1975).

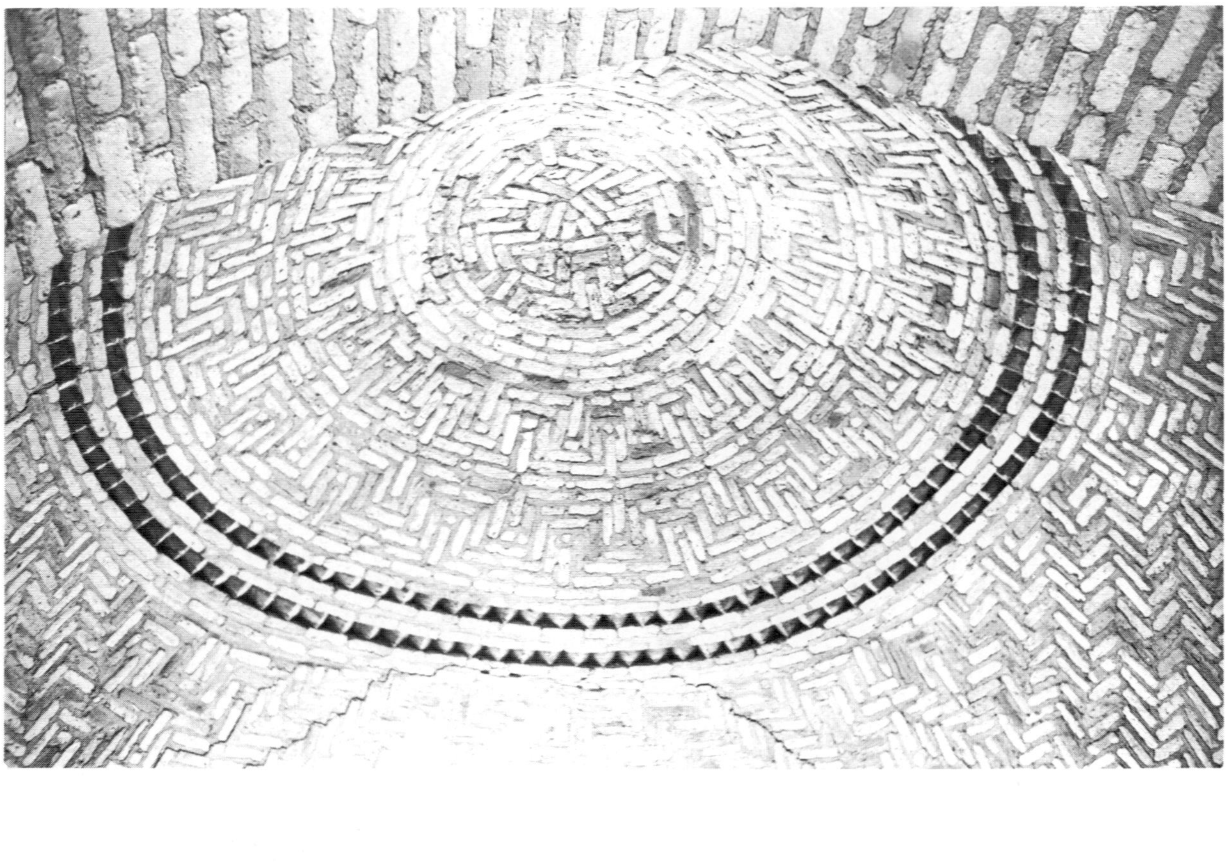

23. Bricks used to finish a vaulted roof (top). A brick kiln in the desert (bottom).

Brickwork: form and decoration

When Sir Edwin Lutyens wrote "Do not speak of Persian brickwork but rather of Persian brick magic" (Lutyens 1933, 118) he did so with the authority of one who was himself a master in the use of that material and who delighted in it. In terms of architectural proportion and sculptural form, particularly in their domed buildings, the Persians have few equals, and the imagination shown in two and three dimensional patterns produced by ingenious brick bonding is marvellous. The type of brickwork which Lutyens had in mind is beyond the scope of this book, but the skills of those who created it spill back into the vernacular where both form and decoration usually spring from a functional need. Examples constantly occur in both mud and baked brick brickwork (as the illustrations throughout this book show). On the other hand, enjoyment alone seems to be the excellent reason for much decoration: for example that of the patterns of projecting and recessed brickwork on the turrets of high garden and village walls or at the top of the shade walls of ice pools (fig 53).

Ventilation also gives scope to the bicklayer. The shadow of the holes shows up in the bright light whether they be the more usual honeycomb patterns, or the small arched openings made by corbelling bricks, or the triangles made by tilting them on their sides (e.g. fig 20). Whole walls are thus given bold contrasting surface patterns.

Stone *(sang)*

Good building stone is available in parts of the Plateau, e.g. the black and white limestones of high quality used at Pasargadae and Persepolis. Dressed stone is little used for vernacular buildings and is rarely seen in comparison with brick. However, travellers have noted stone doors in private houses (e.g. Simpson 1887, 66).

The stone used in vernacular building is usually very rough random rubble. Its use varies according to availability as much as need. In places buildings do not even have stone foundations; in others the whole building may be stone, but this is not common. In some the foundation is carried up as a plinth of random rubble jointed with a lime-mud mortar. As such it protects the wall where it is most vulnerable to wear and tear in a village street, and, more vitally, to undermining by water after heavy rain or snow.

Plaster *(kha-gel & gach)*

Mud and mud brick walls and roofs are plastered inside and out with a mud and straw plaster (*kah-gel*) of a similar mix to that used for mud bricks, but wetter. The resulting rough surface texture of this relatively soft plaster is particularly attractive, the short bits of straw in it being visible on the surface – when it is light enough to distinguish these, the first division of the day has begun for the Zoroastrians of Yazd (Boyce 1971, 129). Dung may also be added to the plaster mix for strength and elasticity when it can be spared from its primary use as fuel, and lime is added to the mix where resistance to damp is considered to be important.

Both gypsum (*gach*) and limestone can be quarried over much of the Plateau and are still burnt in some village kilns. Gypsum has long been used for much of the marvellous stucco decoration for which Persian buildings are famed. Now it is frequently used instead of lime for the hard smooth plaster used internally on burnt brickwork and produces a finish similar to that of the hard modern plasters common in the west.

Carpentry

Napier Malcolm was somewhat unkind about Persian carpentry: "Our cat could generally get in through a bolted door" he wrote of his house in Yazd, and of the carpenter "He never dovetails, and he disguises the inaccuracy of his joints with plentiful deposits of plaster" (Malcolm, 1905, 20 and 23). In the Plateau villages, the carpenter probably is the plasterer and is very likely the bricklayer too. Carpentry cannot develop at the village level without a readily available supply of timber. The village builder makes the best of what he has and is a master of improvisation, re-using any bit of wood again and again. Good carpentry, joinery and cabinet-making are rarely part of the vernacular tradition of the Plateau, so they have been omitted from this book.

Flat roofs

Flat roofs are common wherever there is timber. It is a precious commodity and is often the only material salvaged when a building is abandoned. In villages on the desert fringe there may be enough water for a poplar plantation, a fast-growing crop which produces straight poles, easily stripped of their bark. Date palm oases also provide some timber.

Poles laid as ceiling joists are covered with boarding, brushwood or reed mats fixed with battens. Then comes the usual lime-mud mix which is spread layer upon layer as it dries. Each layer is thoroughly compacted with a stone roller. The total thickness depends on the insulation needed; commonly 50 cm (18 ins) or more. Rainwater channels are formed in the mud roof which carry water off through long wooden spouts projecting well clear of the walls.

The pale colour of the mud has the advantage over darker materials of reflecting some of the sun's radiation. This applies equally to domes and walls. At the same time

25. A flat roof (in need of repair). Layer upon layer of mud is rolled on boarding (or reed-matting) and brushwood, where it is available, to give excellent insulation.

the softness of the colour and texture is easy on the eye.

There are no ceilings in the usual peasant house. The pale gold of the reed matting or boarding and the muted tones of the timber joists are much more comfortable to the eyes after the outdoor glare than a flat plastered ceiling.

Arches, vaults and domes

The vaults and domes seen on the Plateau have amazed Europeans for a long time. It is not their huge number, nor their size nor shape that most astonishes the westerner, but their method of construction. John Fryer recorded " .... and what is more wonderful, only with Mud and Clay, they will rear most spacious Arches, without other matter of Assistance". (Fryer III, 1915, 17). The absence of centering or timber formwork, universally used in the west, fascinated him as it did Tavernier "..... for without all the trouble and time we spend, they will presently raise an Arch so broad and high as easily demonstrates the skill of the Workmen". (Tavernier 1687, 148). Certainly we take "trouble and time" to get proper centering made for the bricklayer or mason; in the west no-one would contemplate building an arch without it. The Persian builder, devoid of timber, contrived to

26. Domed and vaulted structures: Zoroastrian funerary pavilions near Yazd, 1975.

manage differently for much of his work, only using formwork for big spans and for buildings of architectural pretensions.

A bewildering variety of vaults and domes, many of which are hard to describe or define, are to be found right across the Plateau. It is often difficult to imagine how they were constructed, but this must depend on a combination of factors which the bricklayer takes for granted as a result of experience and custom. Indeed, to pause for thought during a vault's construction might prove to be its undoing. Much depends on the speed of work and this is comfortably achieved by men working in small teams using bricks and mortar of a kind which combine to produce a stable structure remarkably quickly. The shape of the vault also aids initial stability since there is a traditional preference for forms which follow that of the line of thrust. This is the line of an inverted catenary curve, in which the height is at least half the width. A catenary is the curve taken by a length of chain or a piece of string, held at each end but hanging free. The best known example is that of the famous vault over the throne room of the Sassanian palace at Ctesiphon. In the villages this is usually formed by eye by the bricklayer, but plaster jigs are also used.

All the bricks used in the typical arch are the same shape as those used in walls; no specially shaped springers

27. (top left) Barrel vaulting. The end of a barrel vault being closed at right angles to the main vault.
28. (bottom left) Barrel vaulting. The end of the vault curves from the end wall.
29. (right) Building a quadrapartite vault.

or key bricks are necessary since the thick joints give the tolerance required by the line of the arch. (The idea of a key stone often seems to be quite unappreciated; indeed a few horizontal bricks have been seen packed into the space where a keystone might be expected).

Both the shape and porosity of the brick are important. The tile-like brick in traditional use presents a large surface area to the adjoining bricks in the arch: 20 x 20 cms (c.8 x 8 ins) which is nearly twice that of a brick of western proportions. The tile shape may have evolved so that it could be used equally easily for walls and arches. The relative porosity of the Persian brick (and particularly of the mud-brick, which is of almost the same mix as the thick mortar used for the joint), allows for strong suction. This suction provides enough sticking power in the joint to allow the arch to hold up without support for the few vital minutes during construction before it is completed. Once finished it is of course structurally sound. The construction of vaults with rectangular western shaped baked bricks is not now usual (fig 27) but this may be accounted for by the use of *gach* mortar, which is very quick setting, and speed of construction.

Various forms of vault are used of which a barrel is one of the most common. It is formed over an oblong room like half a cylinder lengthwise bearing on the room's long walls. In the West a barrel vault is usually made with bricks or stones set vertically in relation to its long axis (like a series of arches built touching each other). But vaults in Iran are often constructed with the rows of bricks lying diagonally instead of standing vertically so that each line of bricks can rest against the "arch" just constructed. The first bricks of the vault bear on the end wall of the room as well as its long walls. This method is also used in parts of Egypt where vaults of up to three metres span are constructed thus (Fathy 1973, 95). If we tried to do this in England with our comparatively dense and heavy bricks with narrow mortar joints the vault would almost certainly slip; horrible scenes might ensue and certainly nothing accurate would be achieved. Not all Persian vaults are now constructed on this principle. Bricks are also set vertically (western fashion) and defy gravity during construction with the help of *gach* mortar as described above. The end of a barrel vault is sometimes completed by vaulting at right angles to the main vault, making a semi-dome.

A rather similar system is used for putting a quadripartite vault over a square compartment. Vaults of up to about five metres across are commonly constructed in this way. Work proceeds across all four corners until the hole in the centre decreases so that there is only room for one man to work. Construction goes rapidly, the bricklayers being fed with bricks tossed up from below, each ring of bricks exerting a thrust on its neighbour as it is completed.

Another method is to arch across the corners of a square space forming squinches to support a dome which bears elsewhere on the walls (fig 31).

30. A hole left for ventilation in the crown of a dome. This is the desert shelter seen in fig 16.

31. Inside the travellers' shelter seen in fig. 16. Squinch arches built across the corners carry the corbelled brick dome. Mud benches and wall alcoves are provided for travellers and their belongings.

Domes may be plastered both internally and externally so the method of construction is not obvious. Some of the bigger ones, for instance those of many of the cisterns and ice-houses (figs 45 and 54), are constructed of corbelled brickwork. The bricks are laid horizontally, each succeeding course projecting a little further inward than the course below it until they eventually converge at the crown of the dome.

In either method of construction the crown may be topped by a ventilator or be filled in with bricks as a small flattish saucer.

A series of vaults (or domes) may not always appear as such from above. Flat roofs are often constructed over them in order to prevent snow (or the rare rain) lodging in the valleys, or to provide a flat roof to sleep on. To save weight, the roof is supported on a series of small arches which rest on the vault. This roofing of the vaults is analagous to the western practice of constructing pitched roofs over the vaulting of a medieval cathedral.

The outward thrust from a vault or arch is usually taken by the thick walls supporting it. Where there is a tendency for walls to spread, tie beams or rods are introduced, usually without any attempt (or hope) of concealment.

Thermal properties of vaults and domes

Domes and vaults lose more heat than a flat roof in the cool of the evening, not only because they have a bigger surface area from which to radiate heat, but because they lose more heat by convection when breezes blow. Even in relatively calm air small changes of pressure caused by air moving over the surface of a dome will speed up the air flow.

Domes which have openings in the crown (*hava kesh*: air draw), give good ventilation to the room below, particularly when the outside air is cooler than that indoors. Air will also be sucked out of the dome when a wind blows and causes low pressure in its lee. (see Bahadori 1978, 15).

Whether or not the dome is ventilated the space within it will gradually fill with warm air from the room, leaving the relatively cool air at floor level, as many people living in high-ceilinged rooms in cold climates know to their cost. Indeed Persians suffer in this way too. The traditional way to keep warm is to sit on the floor with your legs stretched out under a *korsi*. This is an adaption of the ubiquitous low wooden table *(divan)* which is about two metres square and thirty centimetres (one foot) high. A bowl of glowing charcoal is put underneath it and the heat is kept in by a quilt or a carpet which is spread over the top and is large enough to reach the floor on all sides.

The sense of coolness given by the height and proportions of most domed or vaulted rooms is probably as important to the occupant as their actual temperature. As mentioned earlier, relief from the outdoor glare matters as much as relief from the heat and the gentle quality of the indoor light is enhanced by the natural colour of the mud brick or mud plaster which is very like the underside of a young field mushroom.

*31. A ventilator (*hava kesh*) over the entrance hall of a relatively grand village house.*

Planning for the climate

The planning of villages, buildings and the rooms within them is of as much importance as that of the building materials and structural form in countering climatic conditions which would otherwise be intolerable. The arrangement of houses and the way of life of those that use them has been brought to a fine art by the citizens of Yazd, the city on the western edge of the great central desert which is renowned for its windcatchers (see Chapter V where Yazdi houses are described in detail).

The life of a village on the edge of the desert is focussed on its water outlet (Chapter III). Its gardens and cultivation help to act as a buffer between the desert and the village houses, and villages are usually walled to give further protection from the desert wind which all too often resembles a sand-blaster.

The narrow street (sometimes vaulted over) and the covered bazaar, typical of most desert communities, give shade and shelter throughout the year; these, and the use of flat roofs for sleeping, on summer nights, are of equal importance to their comfort.

The aridity of the Plateau means that underground rooms can be used without much trouble where circumstances suggest, usually to combat heat. Problems of damp rarely arise. The most obvious example of underground or partly underground rooms are those used by

32. A shady village street. Fahraj near Yazd (1971).

33. The shaded courtyard of an isolated farm on the plateau – the focus of the farm was a big pool in the middle of the yard.

semi-nomadic people. These makeshift dwellings, dug out of the ground, perhaps with rubble stone walling to retain the soil but with only a rough thatched roof to keep out the weather, make the most of the thermal insulation given by the earth. Persian dwellings of this kind are usually temporary and primitive but in some parts of Iran, large animal shelters are dug out underground to winter sheep so that they will be safe from the cold and the snow.

Shelter is needed in the Plateau villages not only from the weather. Suspicion of the stranger, fear of theft and the Islamic desire for privacy for the women of the household combine with the climate to make buildings inward-looking, so they are often planned around courtyards. Windows are seldom found on outside walls. The street face is blank except for one sturdy door indicating a habitation within.

In bigger houses the street door leads into a small hall or ante-room, often hexagonal or octagonal and domed. From the hall, a passage, probably off axis and thus giving the newcomer no "lead" into the house, links it with the courtyard which is the heart of the house. This passage will probably turn a corner, making the courtyard invisible from the entrance hall as well as the street and thus adding to the sense of privacy.

In the courtyard, trees and a pool of water help to create what is in effect a micro-climate usually secure from the wind, heat and dust of the outside world (see also Chapters III and V). The cooler air will remain sunk in the courtyard below the hotter which lies over the village at roof level. This is particularly marked in two-storied buildings. In bigger village houses the rooms are arranged for summer and winter use according to the sun and shade, the summer rooms are north facing and the winter south. The other walls of the summer rooms are protected from the direct rays of the sun by corridors or stores.

South facing winter rooms are planned so that the lower rays of the sun can shine through French windows which open onto the courtyard, warming not only those using the room but heating the tiled or mud floor which will re-radiate the heat stored in it when the sun moves on.

Many houses have upper rooms (*talar*) either at first floor level or simply raised half a metre or so above the courtyard. They have openings in the long side and d provide a shaded and relatively airy place to sit according to the season. Their roofs are a little above that of the general roof level so they catch the breeze.

The *aivan*, a big, vaulted room or porch open on one of its short sides, plays an important part in Persian building, both by providing a space which is ideal for the indoor-outdoor life led in a hot climate, and by protecting the rooms behind it. It, too, has a higher roof. Most westerners associate aivan with the courtyards of mosques, but in bigger houses two often face each other across the main courtyard one to catch the winter sun and the other, on the shady side, to protect the summer rooms and for summer use.

Typical also of Persian buildings is the "unlinked arcade", a series of smaller porches along the face of a wall. These too are admirably suited to an indoor-outdoor way of life. Perhaps the best examples are those of the traditional caravanserai which has such porches on all four courtyard walls and, often, outside those flanking the entrance (fig 34). Floors are raised half a metre or so above the level of the ground so that travellers and their personal goods could be clear of the jostling animals. Each recess usually has its own fireplace for cooking. Behind is a private room where the member of the caravan who had hired it slept. Bigger aivans off which suites of rooms open, were planned in the middle of each side; these could be hired by richer travellers.

Traditional bazaars are basically made up of deep porches facing narrow streets with rooms running back from the individual arches, each arch being the shop front. Workshops around yards are often designed on the same principle.

Alcoves are a typical and very attractive feature of Iranian interiors. The great thickness of the mud walls means that alcoves are not only easy to incorporate in the design but that they have the asset, from the builders' point of view, of making the walls lighter and using less material. They are particularly useful for stowage of household possessions (bedding, clothing, cooking utensils ....) or the tools of a trade in workshops – rooms which traditionally have little furniture, and which look all the better for their sparseness.

In working buildings or village houses, alcoves tend to occur haphazardly where they are needed, but in some the alcove is carefully positioned to contribute to the architectural effect of the room.

A striking characteristic of some interiors is the sense of lofty spaciousness which is achieved without loss of human scale. It can be found in all kinds of building ranging from the palaces and madrasseh of royal patrons to travellers' shelters by desert roads or in the more important village houses. The common theme is the strong horizontal division, a cornice usually as little as two metres or less (six ft) above the floor, which divides the treatment of the wall surface into two distinct areas. The eye tends to stop at this line, dwelling on the wall surface below it which is made up of a decorative pattern of low doorways and alcoves designed to a very small scale – much smaller than is common in the West.

Above the cornice level there may be another row of niches or the wall may be blank, but often the vault springs from near this level. At all events the considerable volume of the room above the cornice (more than half) makes for coolness and a sense of space while the eye dwells on the small scale below it. Thus lofty rooms, often with a small floor area, which could be opressively vertical in proportion or overpowering in scale, are given

34. (top) A typical caravanserai, constructed in the (late?) 19th century close to the site of the famous medieval caravanserai at Za'farineh near Sabsevar (photo 1970).
35. (bottom left) The entrance door of a house – this example happens to be more important-looking than many. Note the wooden water spout.
36. (bottom right) A vaulted shelter with niches for storing goods.

human (or smaller) scale and horizontal emphasis.

It must be concluded that village building is the result of a strong tradition which has been evolved from the experience of countless generations who have lived close to harsh desert conditions. The result is that the village builder possesses a rare combination of skills of which he is largely unconscious.

It is hoped that this chapter has shown that the vernacular tradition is based on an awareness of micro-climatology, albeit unwittingly, from which is derived an appreciation of the importance of thermal insulation and heat storage which we in the U.K. have only recently acquired. It has also been seen how skill in the construction of arches and vaults transforms buildings which could otherwise be mundane and produces architecture which westerners, with our admiration for "spatial effect" and "sculptural quality" can only envy but are in no position to emulate without calling upon the services of a structural engineer. Ali, Abbas and Akbar have no need to mention these things. The success of the harvest or of the local football team, or another recital from the *Book of Kings* are more fit topics to entertain the stranger.

* * *

This book contains merely an outline account of building materials and construction practised on the Plateau and is inevitably superficial. Practices vary over such a huge area. Hans Wulff's *The Traditional Crafts of Persia* has an authorative section on building materials and on building practices which are fast disappearing. John Norton's UNESCO Reports give insight as well as detailed technical information into the problems involved.

37. A mud hen coop with a stone to close the door.

3 WATER

The problem of existence on the Plateau is still that of survival in conditions of extreme aridity. The Persian talent for making life not merely possible but civilised under harsh conditions tends to conceal this truth from the visitor for whom water may come out of a tap as it does at home.

This description of drought by the 17th century traveller, John Fryer, could hardly be bettered today "... the fiery Element makes the Mountains gape, the Rocks cleft in sunder, the Waters stagnate, to which the Birds with hanging Wing repair to quench their Thirst; for want of which the Herds to low, the Camels cry, the Sheep do bleat, the barren Earth opens wide for Drink, and all things appear calamitous for want of kindly Moisture; in lieu of which, hot Blasts and Showers of Sand infest the purer Air ..." (Fryer II 1912, 173).

It is impossible to over-emphasise the importance of water and its traditional value to the Persian who for hundreds of years has led a settled life in places which would have remained uninhabited elsewhere (e.g. in the empty heart of Australia or large stretches of desert in the U.S.A.). Even in Achaemenian times (c. 500 B.C.) this skill was an essential part of living. "In the *Avesta*, the sacred book of the ancient Persians, irrigation was a good deed in the eyes of Ahura-Mazda ... (and) the Achaemenian kings granted exemption from land-tax for five generations to all who made land cultivatable by irrigation" (Wulff 1966, 245). At the end of the Achaemenian period the cult of Anahita, Goddess of water and fertility, was officially encouraged and water was naturally a vital element in the composition of the Persian garden from that time on. Nearly a thousand years later this belief in the fundamental importance of water was to be reaffirmed by Mohammed: "Do not those who misbelieve us see that .... We made from water every living thing" (*The Koran* XXI, 31). Its ritual importance is not difficult to appreciate: washing is an essential preliminary to prayer.

Land on the Plateau which is not irrigated is worthless for cultivation and that which provides scant grazing provides none in time of drought, when entire flocks may perish. Up to a million sheep and goats may be lost in a bad year (Cambridge History of Iran, I, 1968, 566). Again, little seems to have changed in this respect since Fryer's travels: " ... all along they cry for Rain, both to fill their Water-stores, and to bedew the Earth, which is most miserably parched ..." (Fryer II, 1912, 339)

Indeed the necessity to irrigate and the skill to do so is so rooted in the Iranian mind that the idea of introducing new species of trees or plants which might withstand conditions of extreme cold as well as aridity tend to be received with apathy (e.g. a species of eucalyptus which Australian advisers tried to introduce). Water is synonymous with life: irrigation spells success even where trees are planted in the most unpropitious places.

Much of the Plateau is salt desert so this affects the degree of salinity of water gathered on it and thus the use to which the water is put: whether it is good for drinking, cooking, washing or irrigation. In some regions people can afford to be choosier than in others. A small amount of salinity can be an asset in a hot dry climate where too much salt may be lost from the body through sweating. But travellers might heed Marco Polo's warning: "I will tell you what it is like (near Kerman). For three days he (the traveller) finds no water or as good as none. What water there is is brackish and green as meadow grass and so bitter that no one could bear to drink it. Drink one drop of it and you void your bowels ten times over. It is the same with the salt that is made from it ...." (Marco Polo 1958, 38).

Water is re-used in a recognised sequence, not as in twentieth century London where water is said to be drunk at least twice, but as in the days before sewage received the special treatment of modern plant. Even in temperate climates, water is used and harnessed again and again - for domestic use, for power, and perhaps for irrigation so it is not surprising to find its careful husbanding is much in evidence on the Plateau.

The total average annual rainfall over parts of the Plateau is as low as 126 mm. (5 ins.) and "averages" mean less where rain is apt to come in heavy, rare downpours which can themselves cause disastrous run-offs of surface soil. Even so they help to give a picture of the problem.

Mean annual rainfall

	mm	ins (approx)
Bam	76	(3)
Isfahan	126	(5)
Kerman	203	(8)
Qum	144	(5¾)
Shahrud	135	(5½)
Tabas	91	(3½)
Tehran	223	(8¾)
Yazd	67	(2¾)
Zahidan	117	(6¾)

(Based on Table 5, *Cambridge History of Iran*, Vol. I, 1968) Constrast these figures with those of Rasht (on the Caspian) 1355 mm. (c. 53 ins.), or the U.K., say 508 - 2510 mm. (c. 20 - 100 ins).

Dams

Flash floods from erratic rain are contained by small earth and stone dams, perhaps only a metre high, and the water is channelled to cisterns. Bigger dams may be reinforced with bundles of reeds (as was common in 18th century England) and some, like the Band-i Sistan, were renewed annually.

Historically big irrigation dams were fairly rare on the Plateau where dammed water would have evaporated fast. That at Band-i Amir in Fars is a notable exception. It was built about 960 A.D. but was probably constructed on Achaemenian foundations (Wulff, 1966, 246). Its main purpose was to provide water for the Marvdasht (the plain of Persepolis). The dam has a roadway on top supported by a series of thirteen stone and brick arches. It also powered watermills (see P. 83).

The famous irrigation systems connected with the great rivers are outside the scope of this book as are the big twentieth century dams constructed under the Pahlavi dynasty. These huge structures, designed to collect water for industry and domestic consumption in the cities, have inevitably strained resources traditionally confined to agriculture.

The Watertable

The level of the natural water table varies a good deal on the alluvial fans at the edge of the desert proper. In Isfahan, it was less than five metres below the ground level in the 1960's (Wulff 1966, 256). This, as much as that fickle river, the Zayandeh Rud, on which the city stands, must have accounted for the marvellous green oasis in which it is set (now much obscured by smog resulting from industry and the exhaust of vehicles). Isfahan's rainfall averages only 126 mm (5 ins) which underlines the value of its relatively high water table.

On much of the Plateau the water table is deep down; in Yazd and Kerman for instance it is 50 - 65 metres (150 - 200 foot) below the ground. Here, as on other parts of the Plateau, the immense labour of bringing water to the surface made traditional wells impracticable for anything but small quantities of water. This has led to the *qanat* (P. 34), a most ingenious means of providing water not only for the huge amounts needed for irrigation but also for city water supplies.

Raising water

Raising water from wells, for domestic use or for irrigation, was formerly carried out by oxen, mules or camels driven down an incline, or by using the *saqiyeh*, the so-called Persian Wheel with its endless chain of pots. In medieval times in Sistan this process was apparently partly mechanised by the use of windmills (Gold 1947, 9) and the practice seems to have persisted: "In some parts of the country (Sistan) these mills are adapted to work horizontally for the raising of water, but we did not see any of these" (Bellew 1874, 235). None have been reported for many years. The millwheel was possibly geared to an Archimedean screw, as at the Aden salt-pans until the 1950's, or to a large scoop-wheel, as in Holland and East Anglia, or to a chain of pots. Given the fitful nature of the wind except in the summer, even in Sistan, and the extensive carpentry involved, it can hardly have been a very efficient means of raising water (see also chapter VII). *Noria*, large waterwheels which raise water in buckets impelled by a fast-flowing river, were commented upon at Ahwaz by Muqaddasi and still operate on the river Orontes in Syria and in parts of Afghanistan. The only reference to them on the Plateau is in connection with Band-i Amir (Wiedemann 1970, 289; see also chapter VI). The relatively deep water-table below most of the Plateau made the qanat a more effective proposition in the long run, despite the considerable expense, but recent drilling techniques have made the tapping of the water table below the desert a reality.

It is discouraging to consider the huge damage done by modern bore-holes or deep wells in further lowering the level of the water-table – an ironic situation since most were drilled with the best intentions. By 1970 there were 7,400 such deep wells producing water for irrigation: twenty-five years earlier they were scarcely known (Beaumont 1974, 140). When a bore-hole lowers the water-table in country already short of water, the chances are that the desert will increase. The pattern is something like this. Ground water pumping first makes possible an expansion of cultivatable land but this is followed by a retreat. Pumped wells in the large alluvial fans on the desert fringe cause a rapid fall in the water-table, perhaps a metre in a year, so the yields of the old wells (and qanats) dry up. Cultivated land on the desert edge is abandoned and reverts to desert. Bore-holes also lead to poor farming through over-watering; salts are deposited as the water evaporates and this too leads to more desert encroachment (Beaumont 1977, 763). Over-watering is rare when irrigation depends on manual work.

Snow

The unreliability of rainfall and rivers and the infrequency of wells (because of the low water-table) and springs makes snow the prime source of water in much of Iran.

It is the most important source of water for both irrigation and domestic supply in towns and villages towards the desert fringe near the foot of the mountains. Melting snow feeds seasonal rivers but this water is of far less importance than that which is slowly absorbed by

38. Wells and streams:

(top) A traditional well.

(centre) Collecting water.

(bottom) A Persian wheel in Afghanistan. Mules raise water by rotating a boom which is geared to the bucket wheel. The water spills into a channel at a higher level than that of the main stream. These are now rarely seen on the Plateau.

the permeable rock beneath. This rock is gradually saturated by thawing snow (and to some extent by mountain rain) and becomes a vast underground reservoir, safe from evaporation. Its drawbacks are that it may be a long way from where the water is needed both in terms of its depth and surface distance. The qanat (or *kariz*) overcomes this problem.

(Numerous books and articles have been written on qanats. Anthony Smith's *Blind White Fish* gives a first hand account of those in Persia; Philip Beckett (quoted below) was a member of the expedition.)

Qanats

A *qanat* is a tunnel dug out by hand at a gradient less than that of the land through which it tunnels. It thus brings water from its source underground in the permeable rock near the foot of the mountains to the spot where it is needed. This distance varies from a few miles up to ten or more: twenty miles is considered very long. Discharge also differs greatly and varies with the season. Tehran's qanats have been quoted to have given between 40 and 240 litres per second (Goblot 1979, 42); those in the Varamin plain, to the southeast of the city, from 1 to 327 (Wulff 1968, 99). Figures are inevitably misleading but 220 litres per second is considered high (Beaumont 1974, 421); 100 litres (22 gallons) seems to be thought to be good.

Until about thirty years ago, when piped water was brought from the Karaj dam, Tehran's water was entirely supplied by qanats as was that of Yazd before bore holes were drilled. It has been estimated that Iran had 22,000 qanats (170,000 miles of underground conduit) delivering a total of 522,400 litres of water per second, that is the equivalent of 75% of the discharge of the Euphrates into the surrounding plain (Wulff 1968, 94). It was earlier estimated that twenty million days labour might be expended annually on their maintenance and construction (Noel 1944, 191f).

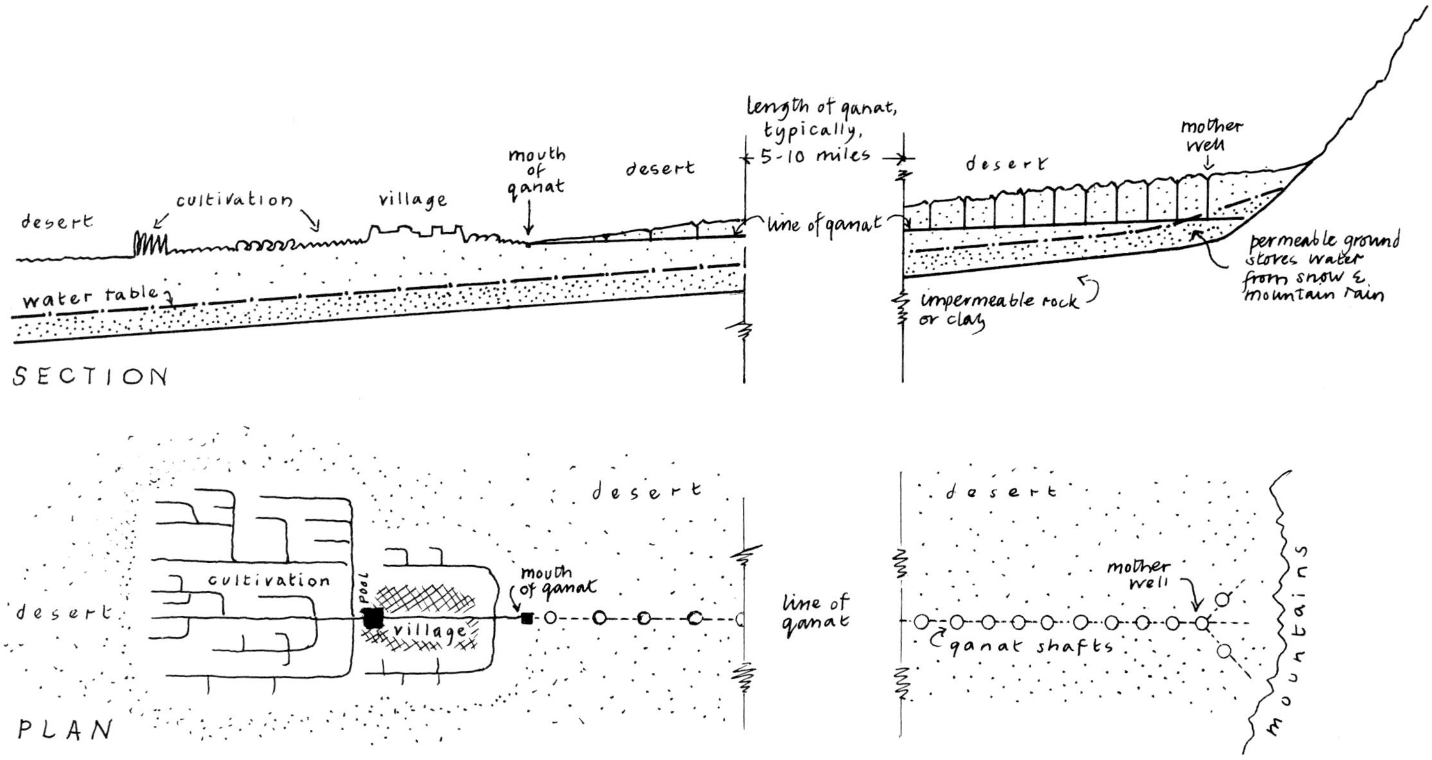

39. (opposite) Man's impact on the vast and lunar desert landscape seems oddly geometric and unrelated to natural form. The qanat and irrigation systems dictate this. The qanats run in straight lines from the water-bearing strata near the foot of the mountains. The line of a qanat is shown on the surface by spoil heaps of earth and stone which ring the shafts.

(above) Diagrammatic plan and section of a qanat. The length of a qanat varies greatly; 2–5 miles is fairly typical, 22 miles is known. The irrigation channels run parallel to and at right-angles to the main channel.

Obviously an underground channel is very much more difficult to construct than a surface channel but the qanat has unique advantages. It taps a vast source of water at a rate of flow which is proportional to the supply. It produces water in manageable quantities so little is wasted and the supply is unlikely to dry up. It brings it to the surface, where it is wanted, by gravity alone. The qanat cannot evaporate. It will not be contaminated and since the water it carries has been filtered by the permeable rock from which it came, it is clean (in some cases the ground through which the qanat tunnels makes it too brackish to drink). A qanat cannot easily be diverted by a thirsty neighbour.

The planning of a new qanat is a complicated business. The speed of the water must be kept below 2 - 3 miles per hour to reduce erosion of the tunnel, therefore the level of the source of water must be related to that of the spot where the qanat is to surface. Various specialists are called in. The first is the water-finder (*muqanni-bashi*). Surprisingly his skills are not like those of the water-diviner or dowser elsewhere who seems to be little known in Iran, probably because he is not searching for a spring of living water which is the object of the water-diviner, but for the equivalent of a vast underground reservoir of water-permeated rock or gravel. The water-finder relies on a combination of minute observation and experience. He carefully examines the sloping land above the desert near the foot of the mountain range for traces of vegetation which have resulted from very slight dampness. It is esential that this relative dampness is not caused by an isolated bed of permeable rock which will dry up because it is only fed by local rain. A qanat must be fed by a big catchment of water in a large area of permeable rock fed by slowly melting snow in spring. Trial wells are dug to check the rate of seepage and the level at which this occurs. There must not only be enough water to justify the huge labour of digging the qanat but it must be at a sensible depth to keep the cost within economic bounds: the depth is clearly related to the urgency of the enterprise.

40. (top left) The top of a qanat shaft. This shaft is ringed by a relatively small quantity of earth and stone which suggests that the qanat is not deep down. The windlass has been left in position by the maintenance gang.

(top right) Working the windlass.

*(bottom) Qanat workers (*muquanni*). The clothes of another muqanni, working below, are neatly folded in the foreground of the picture. The excavated stones are unusual; they seem to have been arranged thus for the sheer pleasure of contriving a pattern.*

'1. *A crumbling qanat tunnel near its entrance (photo John Donat).*

The successful trial shaft becomes the mother-well (*madar-chah*) – between 15 and 25 meters (50–80ft) is considered to be a reasonable depth, 90 meters (300ft) a maximum. However there are three famous wells at Gonabad, each said to be 300 meters (979ft) deep, which were dug in the 13th century; each well steps down in three stages. The qanats which they feed are over 22 miles (35km.) long – their great length being dictated by the depth of the well (Goblot 1979, 35ff). 2 – 5 miles is a more typical length.

Next a surveyor will supervise the digging of guide shafts at two to three hundred metre intervals to give the line and depth of the qanat (Beckett 1953, 48). The qanat diggers (*muqanni*) are then left to get on with their arduous tunnelling. Work in this ancient and skilled craft often runs in families, fathers teaching sons, blind members included. If the qanat diggers do not wish to dig on a certain day (an omen might deter them) no-one will encourage them to work. The qanat has not been given the name of the *khooni* (murderer) for nothing. The booming made by loose stones hurtling down the shaft is frightening enough, but the dangers of being trapped by a collapsed tunnel or suffocated are worse.

The diggers tunnel like moles bringing the soil up the vertical shafts by means of bucket and windlass (figs 40) thus the course of a qanat is punctuated by what looks like a line of giant mole-hills or bomb craters at intervals of about 50 metres, more closely spaced in the deeper qanats since the shafts are used for ventilation as well as construction.

Very deep qanats may be ventilated by pairs of shafts side by side. A lamp is placed at the bottom of one shaft, causing the air warmed in it to rise creating a downdraught in its partner (Beckett 1953, 48).

Most diggers want to work in a dry tunnel for as long as possible so the tunnel is begun at the outlet and is dug back to a point just short of the mother well. The line is taken by sighting, using lamps within the tunnel and by listening for the sound of fellow diggers in the vertical shaft ahead. The depth is checked by dropping plumb-lines marked with knots down the shafts.

Eventually when the breakthrough to the mother-well is made there is considerable danger of drowning. Usually the qanat runs back some distance beyond the mother-well into the permeable rock in order to supplement the water absorbed by the mother-well itself, or several short spurs of qanat may be dug back to help to feed it.

The tunnel is usually c. 1.20 metres (4 feet) high and c. 75 - 90 centimetres (2½ - 3 feet) across. In loose soil, elliptical clay hoops c. 109 x 63 cms (43 ins x 23 ins) are used to reinforce it. Their weight, c. 22 - 25 kgs (50 - 56 lbs) is important since they have to be carried some distance in the confines of a tunnel. These hoops were often baked in a field oven on the spot but are now usually mass produced in a village kiln.

Qanats are cleaned annually but after heavy rain maintenance is usually essential as it certainly is after an earthquake. When a channel cannot be cleared the spot is bypassed: this too can be recognised on the surface by the deviation of the line of "mole hills".

The origins of the qanat

The word qanat, now generally used in Persia, is Arabic; it means lance or conduit. The original Persian word is *kariz* and this is still used in Afghanistan. Farsi (the Persian language) is particularly rich in words connected with qanats (Humlum, 1964, 92). They are generally thought to have originated in northwest Persia (ancient Armenia), part of which is a rich mining area. It is likely that they were first developed by miners who were attempting to drain their workings. Indeed it is difficult to say whether a near horizontal adit has been dug primarily as a passage to connect vertical shafts, or to drain a working, or to carry water (as do qanats); or to increase the output of a well. The Egyptians, from 1500 B.C., drove horizontal passages into the strata at the bottom of wells to this end, (Forbes I 1955, 146), but the use of qanats in Egypt dates from the Persian occupation.

By this time, 525 - 322 B.C., they were widely used; it is thought that Darius's capital, Istakhr, may have been supplied by qanats (English, 1968, 175). They were common in north India in 300 B.C. and spread to Trans-Caucasia, Afghanistan and Chinese Turkestan. The Parthians (223 B.C. - 226 A.D.) were to increase their use in Persia, and with the coming of Islam to Iran (642 A.D.) qanats eventually spread, via North Africa and Spain to the New World where new ones are still constructed.

The influence of qanats on the settlement of the Plateau

Clearly an appreciation of the qanat system, both its potential and its limitations, is essential to the appreciation of the overall pattern of settlement on the Plateau. It accounts for flourishing settlements many miles from any visible source of water, and for their collapse and abandonment if the life-giving qanat has been cut either by natural disaster or perhaps the hostility of an enemy.

Qanats also tend to dictate the social pattern of settlement within a town or village. In a perfect settlement all the drinking water would be collected near the mouth of the qanat before it is distributed in open channels (*jūb*); in some it is, but in many the water in these channels is also drunk downstream from say, the clothes washing place, or below where the animals are watered. Thus the pure water supplied by the qanat becomes what has been described only too aptly as the means of "the simplest distribution of infection known to man" (Smith 1953, 76).

Thus qanats influence the social status of different parts of a town or village which must depend to a large extend upon the quality of the water running through it: it is clearly more desirable to live near the mouth of the qanat where the water is relatively pure,while that in the open water channels which it feeds becomes increasingly contaminated. South Tehran for instance, around the bazaar, once the heart of the city, came to be considered to be much less desirable than to the north (towards the Alborz mountains) where the water is naturally cleaner, being nearer the mouth of the qanats. Thus the city has spread northwards to join the smart suburbs which sprang up around villages along the lower slopes of the mountains (e.g. Gulhak) with the coming of the motorcar. They also enjoyed the greater height.

We have a similar pattern in London and other cities which once suffered smoke pollution created by the prevailing wind. The southwesterlies consistently blew smoke eastwards over the town, hence the West End crept further west leaving the East End to become more polluted and less desirable.

Law and custom

Water and pasture were considered to belong to all Muslims in common; this was presumably a relic of the life of a desert community. A book of laws called the *Kitabi Quni*, said to be written in the 9th century A.D. (and certainly extant in the 11th), governed the qanat. By law water cannot be bought or sold. Traditionally such sale was forbidden by the Prophet but permission for its use had to be obtained from the owner of the spring, well, or qanat from which it came.

The owner of a qanat owns not only the land directly over the tunnel but also that which is necessary for its full enjoyment. The borders (*harim*) of a spring or a qanat are 500 *gaz* (approximately 546 yards) on all sides in soft soil and 250 *gaz* in hard soil. No-one may dig within another's borders (Lambton, 1953, 16, 217, 210, 199).

The ownership of a qanat becomes increasingly complicated over the years through inheritance and marriage. The use of the water it provides is most carefully controlled. Its measurement is based on time and is gauged by the amount of water that will flow through an aperture in that time. For instance, the single channel leading from the mouth of the qanat might be led to a pierced stone slab (*maqsam*) which divided it into, say, five streams. These would lead water off to different parts of the village land for a stated time. Until recently (and perhaps the custom persists) this was timed by floating a bowl which had a hole in it, until it sank. Each time it sank a stone, used as a counter, would be moved. A plot of land would be due for so many stones (*sang*) worth of water. When time is up, sluices (or simply shovelfuls of earth) divert the water to another plot. This makes for interest when walking about a Persian village since a considerable channel of water may unexpectedly cut a route which shortly before was dry. That some of this *ad hoc* digging goes on surreptitiously after nightfall is not surprising.

The illegal diversion of village water channels, often tunnelling under mudbrick walls to achieve the desired direction, are all reflections of concern for one's land and what grows on it. Without water it will be useless. The deep concern for his water rights, common to all Persians, is a direct expression of the instinctive proprietary right we all feel for our own territory, be it a backyard or something more extensive. The peasant's plot or the landlord's fields revert to desert without constant irrigation. The builder of a qanat must be rich but since his land is worthless without water he is behaving as might any landlord in making his land profitable. In other circumstance it might be necessary to drain or to dam to achieve the same ends. These ends matter to everyone and always will. Their importance was nicely emphasised by the steward of a large estate (near Hamadan) who found that nearly all instances of violent crime of which he heard were connected with irrigation (Forbes-Leith 1927, 47).

Cisterns *(āb ambār)*

In cities, water is usually stored underground, sometimes in big columned cisterns, but more often in smaller tanks within the buildings, each house having its own which is fed, as its turn comes, by a branch from a qanat.

In and around the Plateau villages the free-standing domed cistern is typical; these are constructed either to store rain water or to be fed by qanats. The cistern is dug 5 or 6 metres (15 or 20 feet) into the ground which is only retained by masonry if it is crumbly. It is domed or vaulted over to reduce evaporation and to prevent if filling with sand and dirt. The dome springs from a low wall (a metre or less high) so it looks like the top part of an egg in an egg-cup. It is often constructed of corbelled brickwork and usually has a hole in the crown for ventilation. Sizes vary according to the water supply: a diameter of about 6 metres (20 feet) is common but cisterns may be 15 metres (50 feet) or more across. Both the dome and cistern are lined with waterproof plaster.

Traditionally the low wall is built of rough stone which inevitably has an uneven top. This is finished off with about half a metre of well consolidated mud which can easily be brought to a good level on which the mudbrick dome can sit.

Occasionally cisterns have stone domes or vaults. Thévenot recorded one such, "four to five fathoms in diameter" and "very deep" (Lovell 1687, 128). Recently an unusual stone structure was seen in Lar: the outside of the dome, which was about 10 metres in diameter, appears to have six rings of projecting "steps", thirty-six "steps" in the lowest ring and twelve in the top, making a curious prickly pattern in low relief; they were probably intended primarily for maintenance.

Not all cisterns are circular: some are rectangular and barrel vaulted. One surveyed was vaulted by a series of curving "barrels" which made it look like a giant caterpillar, 15 metres long (fig 43). Fryer recorded a great variety of plans "some Spherical, others Transverse, cutting one another in manner of a Cross; others, and the largest Oblong, Square, Orbicular or Oval" (Fryer II 1912, 169).

The main advantage of digging cisterns into the ground, a practise which denies the use of gravity to distribute the water, is to keep the water cool. Persians tend to be more particular about the temperature of the water they drink than its purity, a condition with which their own antibodies can to some extent deal. It is natural to relate coolness with purity: warm water tastes dead. A shepherd-boy, having asked for water from a plastic bottle, had no compunction about spitting it out when he found it was tepid, despite the fact that it was in short supply and there was none other to be had. The water would have been cool had it been kept in the skin or unglazed pitcher to which the boy was accustomed, so this untypical lack of poise may have been caused by fear that it was bad. Conversely confidence is often misplaced since both large volumes of water and moving water are traditionally thought to be good sources for drinking

The quality of water in cisterns which are surface fed from whatever source varies considerably. The water is usually drawn by bucket through small openings at ground level. Inevitably, sand and much else tends to find its way through these openings. Most of this naturally settles and bacterial actions to some extent purifies the water as in a modern septic tank (water discharged from a septic tank should be pure). The bacterial action

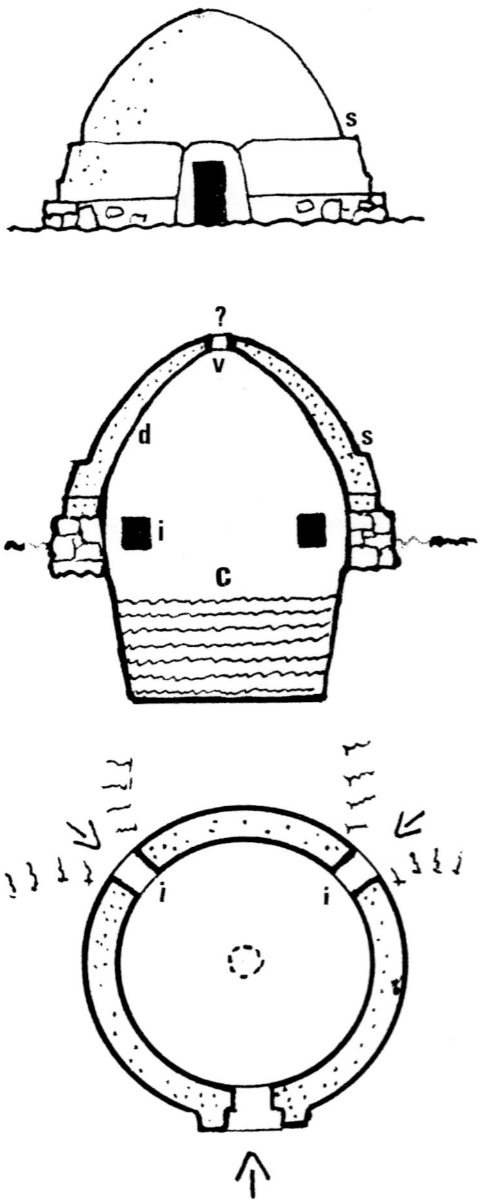

c cistern
d dome
i inlet for water
s maintenance step
v ventilation hole

42. (above) A cistern from which water is drawn at ground level by bucket. Such cisterns are usually fed by surface water.

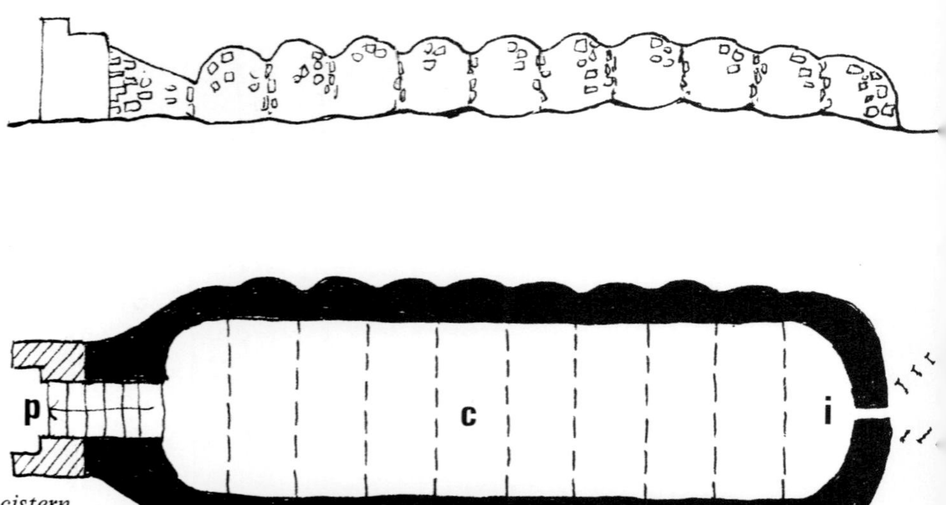

43. (right) A cistern fed by surface water, roofed by plastered stone barrel vaults which are supported by brick arches. A flight of steps leads from the brick porch to the water.

c cistern
p porch
i inlet for water

is dependent upon a supply of fresh air which comes through the openings in the wall as well as the crown of the dome; and also on a regular change in the water, but this is less predictable. The condition of water drawn from the cistern must also depend both on the level in the cistern from which it is taken and on whether the bacteria have been able to tackle its contents: a drowned donkey could take a long time to forget. A bucket breaking the water's surface will be inclined to collect slime, unless the water is drawn with great skill; it will also disturb the bacteria which are busy at work. When cisterns are drained and cleaned out the residue is carefully conserved since it is a valuable fertiliser (Siroux 1949, 128).

Water loses heat by surface evaporation particularly when a wind blows through the ground level openings. In those cisterns with ventilated domes the warm air can escape as it rises,thus creating air circulation.

The water in a qanat fed cistern comes in cool and clean and should remain so because, unlike the surface fed cistern, the qanat fed is often sealed at ground level where the dirt and foreign bodies tend to get in. The water is drawn from a tap near the base of the cistern deep down below the ground. A long flight of steps, covered by vaulting, leads down to the tap (fig. 46). Here a massive wall about two metres (c.6 feet) thick takes the outward thrust of the water where the earth has been cut away. This is a wonderfully cool place in the heat of the day. The staircase has to be straight to allow for natural lighting but it is shaded by a porch at the top of the flight. This porch is reminiscent of an *aiwan* in a mosque; both share a sense of sanctity and sanctuary.

The best cisterns are ventilated by wind-catchers (see Chapter V). Two or four are usual; six towers adorn one of the most prestigious cisterns in Yazd (fig 47) The water is further cooled by the evaporation caused by

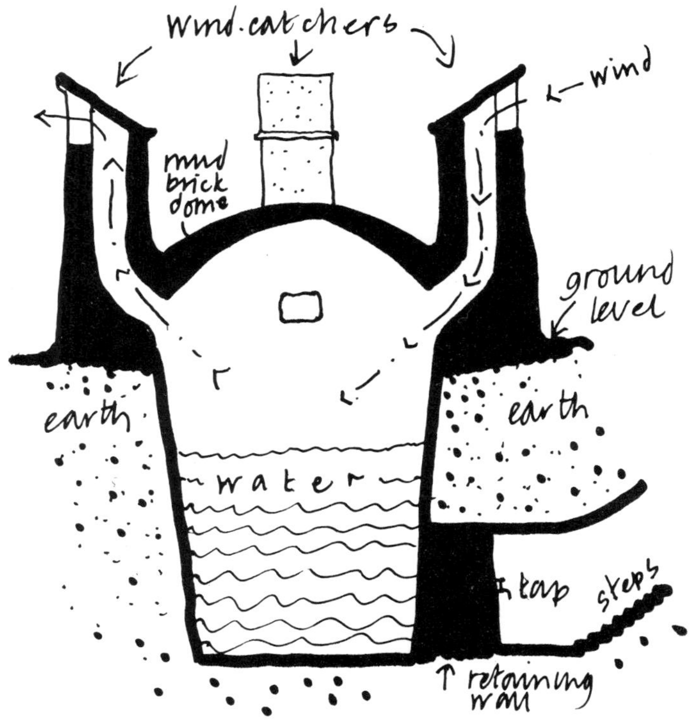

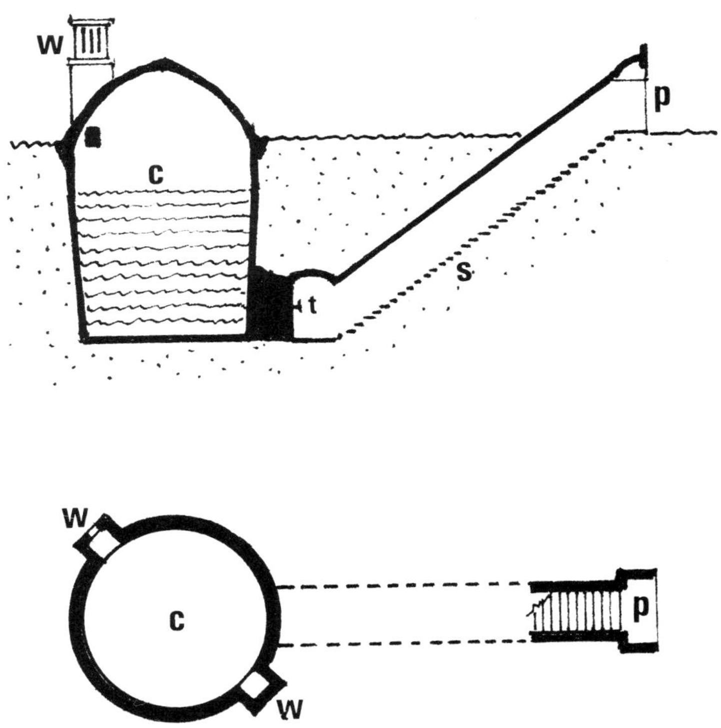

c cistern
p porch
s steps
t tap
w wind-catcher

0 10 m

44. Cisterns ventilated by wind-catchers. Most such cisterns are fed by qanats and are usually ventilated by wind-catchers.

45. A closer view of a similar cistern to fig 46 which is also ventilated by windcatchers (near Yazd, 1961).

46. A similar cistern to that opposite. The porch on the left protects the under-ground flight of steps which leads down to the tap near the bottom of the cistern on the right. Two lines of qanats can be seen in this picture (top).
Steps down to the tap near the base of the cistern (bottom left).
47. A cistern in Yazd with six wind-catchers, 1975 (bottom right).

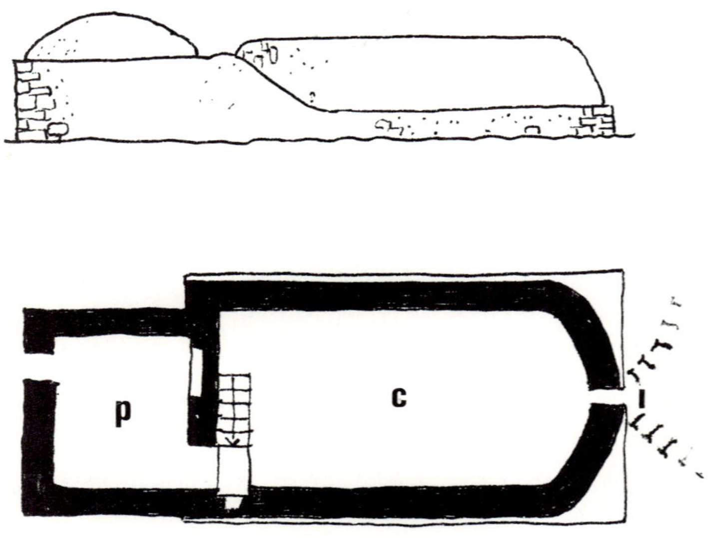

the dry (and often warm) air drawn across its surface; there is likely to be a difference of up to 20°C (35°F) between the water temperature and the ambient mid-day summer temperature. The water is at a perfect temperature to drink. This cannot always be claimed for that from a modern refrigerator which tends to be too cold for someone who is hot and thirsty.

It is hardly surprising that accounts of cistern water vary. Today, wherever possible, drinking water comes from qanats or springs (if not from the piped mains supply), but the seventeenth century traveller often encountered cisterns fed by rainwater. Thévenot found cistern water " ... very unwholesome, and it is good to quench a red hot iron in it, and Strain it through a Cloath, because of the worms that breed there in" (Lovell 1687, 132). Fryer travelled among Persians who were suspicious of water "where the Sun has little influence" but then he was presumably well away from those areas fed by qanats. His experience of caravanserais seem to have been mixed. In one, "Strapping Sunburnt Lassies with little more Cloaths on than a dark coloured Smock, or Frock, bring us Water to Drink out of their Tanks, newly filled, which was Thick, Troubled and Slimy" but he also describes "filthy" caravanserais as having tanks and cisterns which were "commonly very Fine and Costly". (Fryer II 1912, 200, 185 and 178 - 9).

Thomas Herbert, travelling with the ill-fated Ambassador's caravan in Southern Persia in 1627, was favourably impressed. "The arch that covers it (the cistern) is well-nigh equal to the depth, and so well plastered that, when filled by the beneficial rains, it preserves it sweet to the last bucket – which is strange, considering how long it is kept without motion save what it has when the water is drawn out ..." (Foster 1928, 52).

Even small cisterns built along the caravan routes, fed by flash floods, were not to be despised (nor perhaps relied on) before the introduction of motor transport. Now most of these are abandoned and can easily be mistaken for travellers' shelters (figs 43 and 48). It is only after a careful look at the lie of the land that the long earth dams are spotted. These and the low floor level (often filling with sand and rubbish) make it apparent that they were once cisterns.

48. (top) Plan and elevation of a small barrel-vaulted cistern with a domed porch found near a desert road in 1975. (bottom) A ruined cistern with a central entrance easily mistaken for a desert shelter.

The not uncommon sight of groups of cisterns in varying states of repair mystified Fryer as it has modern travellers, but he seems to have hit on the likely reason: "These Cisterns or Storehouses for Rain .... being once finished .... have no Endowment to maintain them, either to keep them clean, or from falling to Decay: .... on which account it is, that about their great Cities so many of all sorts are found, newly built, superannuated, defiled (which they esteem so, if either Man or Beast have dropped in and been drowned) unfrequented, and full of Nastiness; so pervicaciously Vainglorious, that they will have the Repute of an entire Founder, or none". (Fryer II 1912, 169). Clearly problems of endowment and maintenance are neither new nor local, but the cost of traditional building is comparatively economic in Iran.

This may well account for six large cisterns seen outside Khor in 1975, two of which seemed to be new, the others decaying. Another six almost touching , included that already mentioned in Lar. Certainly it is likely to be more hygenic to build a new cistern than to clean out an infected mudbrick building.

Humidity and comfort

Humidity as well as temperature affects personal comfort. Very humid or very dry conditions may be intolerable but the human body is remarkably adaptable in intermediate ranges.

An acceptable humidity next to the skin can be created to a large extent by clothing (a personal microclimate which is carried around). On the Plateau the village men are covered to about the same degree as the villagers of Southern Europe rather than in the all enveloping dress of the bedouin (which is in effect a tent that prevents the dry wind de-hydrating him) but the villager is not subject to unrelieved periods of aridity as is the desert nomad. A woman's traditional over-dress, a sheet of cotton which does not cover her face unless she wishes, speaks for itself. It is called a *chador* which means tent. Many westernised Persian women keep a chador to wear while travelling.

Humidity is measured as a percentage which is known as the "relative humidity of the air" (RH), that is the amount of water vapour actually present expressed as a percentage of the amount it could hold *at the same temperature*. The lower the RH the more thirstily will air evaporate moisture from any source available. An RH of 100% means it is raining.

Since RH percentages are dependent on temperature they do not indicate the amount of moisture present in the air unless the temperature is constant. Disregard of this fact can lead to a misinterpretation of RH percentages in common sense terms of comfort. For instance, in cool climates (e.g. the U.K.) people are generally comforable at a temperature of 22°C (70°F) and an RH of 50% when lightly dressed. On the Persian Gulf the same RH can be very uncomfortable at a temperature of 38°C (100°F) because enough heat cannot be lost by sweating. However, the same high temperature (38°C) is much more tolerable on the Plateau where an RH of only 20-30% might be experienced in a desert town at midday in the summer; in such dry conditions people can lose heat by evaporation as they sweat. In hot, still weather, when the air temperature is considerably higher than that of the body, evaporation does not occur quickly enough and hot stickiness results. But in hot, dry, windy climates our sweat glands cannot keep pace with the demands made upon them by the air, even by continual drinking. A nice balance must be achieved and the Persians have a remarkable talent for achieving it. Draughts of drier (maybe warmer) air are introduced to rooms to help to cool those using them but not necessarily to cool the air in the rooms themselves (see Chapter V).

Water channels and pools, indoors and out, have long been a feature of the best Persian houses. The open water tank is ubiquitous. It is to be found in countless courtyards in Plateau towns and villages. It is nearly always rectangular and often shaded by trees. These tanks are simply humbler versions of the pool in front of the royal pavilion of the Chehel Sitoun, in Isfahan – perhaps the grandest of all garden tanks. Their function is not only to provide water, but to create a micro-climate.

Water brings peace and stillness and the movement of changing light. It also brings contrast from the parched grittiness of the desert. Evaporation from a tank increases the humidity of the air and lowers the temperature in its immediate surroundings. Walls give protection from the wind and blowing dust and a haven of denser, cooler air is created within the courtyard. Dust blowing around will settle as the air cools over the water.

The foliage of plants and trees grown in the courtyard creates a miniature oasis in the heart of the house. Foliage, having absorbed radiant heat from the sun, does not radiate it back in anything like the quantity radiated by paving or dry earth (even in the U.K. the difference experienced when stepping from a green lawn to a paved terrace can demonstrate this). The custom of sprinkling or hosing the courtyard paving with water in the late afternoon and early evening gives added relief and so does the watering of plants. The air is also slightly humidified by the transpiration of water from the foliage.

Thus the sum of many small factors makes for better living in a very hot dry climate: this is not the same as saying that a change in the relative humidity of X% is the cause of the increase in comfort.

Water for pleasure

So far this chapter has concentrated on the more practical aspects of the acquisition and use of water. The Persians do not stop here; they spend almost as much

49. The Pol-e Khaju, Isfahan, built 1660.

imagination and energy in making water positively enjoyable. This they have done at least from the time of Cyrus the Great and probably long before. In his capital, Pasargadae (c. 560 B.C.), for instance, water was an essential ingredient in the creation of a paradise. The great walled enclosure was in essence a walled camp site, not a city as we know it. The King had his Palace but there were no houses. The Court, in the best nomadic tradition, pitched their tents in the shade of the trees, which depended on an extensive irrigation system. In the heart of this park, the pattern and form of the Royal Garden was defined by narrow water channels made in carefully cut stone. These run for over 1,000 metres; they are interrupted at regular intervals by small square pools, each carved from one piece of stone (Stronach 1978, 107). Some are fitted with sluices which would have created the sound of water trickling over little weirs. Such details must have seemed important to the man who had conquered most of the known world but who chose to live when he could at home on the Plateau.

The Pol-e Khaju in Isfahan (1660 A.D.) is a marvellous example of a bridge designed as much for the enjoyment of water as the means of crossing it. It is two-tier with a carriage-way and arcaded walk-way built over a weir which dams the water when it is low. A second walk-way under the bridge crosses just above water level;

50. A picnic on a journey.

the sluices are spanned by stone slabs. Thus the sound of the water rushing or trickling over the weir can be enjoyed in the shade or in the sun by the populace. Projecting pavilions were built overlooking the river, the two biggest in the centre of the bridge being for the Shah's use. Boys dive from the buttresses when there is enough water and on any Friday hundreds of Persians spread their carpets there and light their samovars for that favourite outing, the family picnic. The Persians are perhaps even keener picnickers than the British, and water wherever possible is the focal point of great Friday gatherings. But the humble jub is not discounted, be it for a picnic party or merely a short break in a long journey. The trickle of water and the small patch of shade given by the struggling sapling beside it are eagerly sought. Once found, little kettles are soon bubbling over camel thorn fires and small glasses of tea gratefully sipped.

Appreciation of water was not always so basic. Fryer describes the refinements in the palaces of the seventeenth century potentates " ... Acqueducts and Fountains in the very Rooms below Stairs, are mightily prized, with Water-holes playing to cool the burning Particles of the incensed Air and the whole structure is constituted so, that it may receive these Refreshments every way it is capable ... [they] are constantly sprinkling Water around their seats from these Currents "(Fyer III, 1915, 16f).

Countless other examples showing the importance of water to Persian life will spring to mind but its relevance happens to be nicely expressed by its place in the dictionary where, not surprisingly, it is the first word, and the root of several significant words:

ab, water ....

abad, inhabited, cultivated, populous, flourishing; a suffix denoting place of abode.

abadi, village; flourishing, populous condition.

And some lines later:

abru, *aberu*, good name, reputation (literally: water of one's face).

51. The entrance to a farm on the Plateau. The pool inside the courtyard is fed by the water channel in fig 50.

4 - ICE - HOUSES

The popular use of ice and snow for cooling drinks and food in Persia was reported by several European travellers in the 17th century. "They mightily covet cool things to the Palat wherefore they mix snow or dissolve ice in their Water, Wine or Sherbets" Fryer recorded of his stay in Shiraz, and later of Isfahan " ... The Poor, have they but a Penny in the World, the one half will go for Bread and dried Grapes, or Buttermilk, and the other for Snow and Tobacco ...". He described hunting expeditions on which they took "*Yogdans*(ice-chests) for provisions" (Fryer III 1915, 135 and II 1912, 48). Sir John Chardin was surprised to find that although "The Cold is dry and penetrating more than it is in any part of *France* or *England*, yet the greatest part of the People drink with Ice as well in Winter as in Summer" (Chardin 1927, 239).

Storage and use

Thomas Herbert's (1628) description of Isfahan has a tantalising throw-away line which remains unexplained. Having described the town as being "of no great strength" he goes on " ... towards the outside of the city, a large castle; unflanked but moated about; and several houses within, which guard the treasure, arms and ice there stored". (Herbert 1928, 127). This certainly suggests that ice was precious even if it throws no light on the method of storage.

Fryer reported that the ice was stored in "Repositories" (outside Shiraz) which unfortunately he merely describes as "fine buildings". These were probably similar to the huge domed pits still to be seen in parts of Iran, which were in common use until the introduction of modern refrigeration and road transport after World War II. Only one such pit seen on this survey appeared to be in use (at Za'farineh, near Sabsevar, in 1970; fig 52) but

52. An ice-house near Sabsevar (at Za'farineh), in use in 1970. The ziggurat-like form results from the spiral step which is designed to give easy access to the dome for maintenance.

it was sealed, so we could not go inside it. The doorway of an ice-house is bricked up until the ice is needed.

The practice of storing ice was probably long established in Persia by the 17th century. It may have been introduced as a result of the 13th century Mongol conquest which brought China and Persia into closer relationship. Ice-houses were known as early as the 8th century B.C. in China; these were probably small thatched buildings like their successors there.

In Britain the idea of keeping ice in an ice-house does not seem to have been introduced until the 17th century, possibly by Charles II on his return from exile. Its popularity among the landed classes spread and it became widely accepted as a necessary adjunct to a country estate. Many ice-houses survive, but being dug into the ground or covered with earth for insulation, most are overgrown with trees and shrubs and are not noticeable; thatched examples, like that at Holkham Hall in Norfolk are the exception. Henry Hoare, who had an ice-house at Stourhead, regarded "Ice in Summer" to be "one of the fruits of Industry and application to Business" listing it with "Temples, Grottos, Bridges, Rocks and Exotick Pines" as "the envy of the indolent" – a rather different point of view to that then prevalent in the Persian bazaars. Ice in Britain was not considered to be a common necessity for a long time and it was certainly not for the poor, but in Persia its popularity persisted. An Englishman, Dr. C.J. Wills, who worked with the Indo-European Telegraph Department 1866 - 81, was struck by its ready availability: "A great thing in such a warm place as Shiraz is the cheapness of ice; for about fifteen shillings in dear years and five in cheap ones, ice can be obtained all through the warm weather .... A huge block is thrown down in one's doorway each morning by the ice-seller ...." and "So common is the use of ice that the poorest are enabled to have it, a big bit being sold for a farthing, and even the bowls of water for gratuitous drinking at the shop doors are cooled by it".

Wills continues: "The Persians well understand the art of making water-ices and ice-creams, and various ices unknown to us are made by them, as tamarind juice, pomegranate and cherry-water ices; iced "mast" or curdled milk, and various ices of pounded fruits, as apricots and cherries, which are very good." The quantity consumed, or anyway offered, amazed Wills, when he went to call upon the *muschir* "four conical ices, the size and shape of an ordinary sugar-loaf, were placed in handsome Chinese porcelain basins before each of us" (Wills 1891, 240, 241).

53. An ice-house with its shade wall near Yazd, 1975. The photograph is taken from the south so that the ice-pool is on the other side of the wall. A walkway runs along the wall below the decorative brick parapet. Access steps up the dome can be seen above the porch which protects the entrance to the ice-pit.

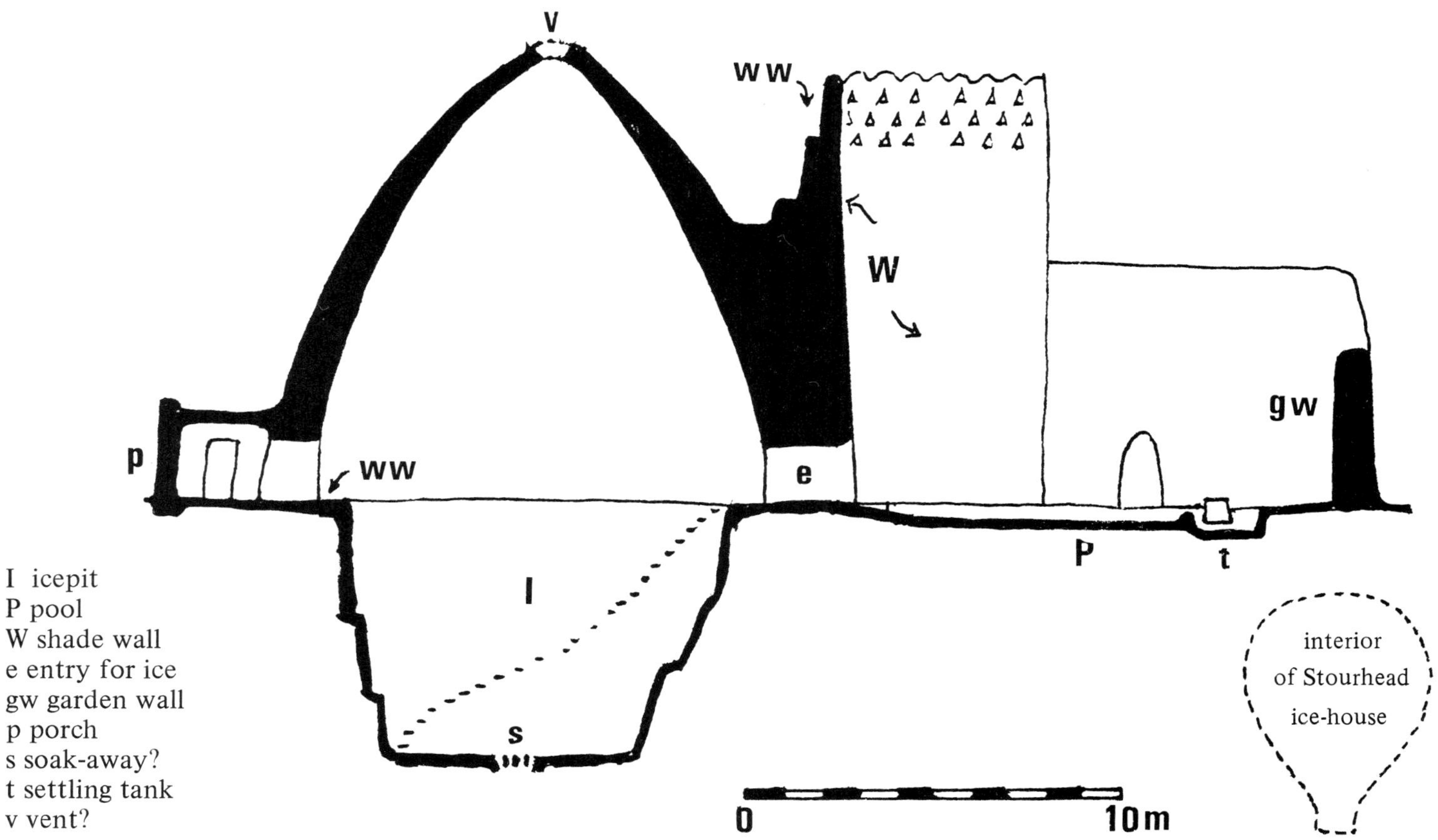

54. Section of the ice-house near Yazd. The size of this ice-house is fairly typical; the interior profile of a good English example (Stourhead, Wiltshire) is drawn for comparison. The Yazd ice-house was dis-used so was becomming ruinous. The corbelled brick dome seems to have had ventilation at the crown and there may have been a soakaway in the floor. The wall and pool were in fair condition. The pool was fed by water from a surface channel which ran through a small settling tank before entering it.

Ice used in such quantity needed considerable storage space. The great demand led to structures of monumental scale and size. (fig 54) shows the Stourhead ice-house (diameter 5 metres, c. 16 feet), one of the larger English examples, drawn to the same scale as a typical Iranian ice-house at Yazd (diameter 12 metres, c.30 feet). However, the principles governing the design of each is the same: the ice had to be insulated and kept dry. The differing climates made insulation a far greater problem on the fiercely hot but very dry Plateau, and drainage of prime importance in damp, temperate Britain.

The drainage of ice-pits was vital to the storage of the ice. Probably most Persian pits had some sort of soakaway in the bottom. It is mentioned by one or two travellers (including Siroux 1949, 132) but it is a difficult point to check because disused ice-houses soon begin to fill with sand or rubbish. On the other hand seepage through the sandy ground into which the pit was dug may itself have given efficient drainage of melt water.

The mud brick vaults and domes covering the ice-pits gave good insulation (Pp. 14,26). The great height of the domes meant that there was plenty of space above the packed ice into which the relatively warmer air could rise, leaving the cold air lying immediately above it. The humidity as well as the temperature of this air is important to the preservation of the ice: it should be kept as dry as possible. The superiority of ventilated ice-houses is described in a letter to the *Gardeners' Chronicle*, 1849 " .... it is the vapour arising from the slow wasting of the ice that is the real cause of it melting fast" (Beaton 1849, 21). The writer, a gardener in the U.K., had designed very successful ice-houses with ventilators. The moisture content of the air in a sealed ice-house, as was common in Britain, could rise considerably over the months. In Persia some ice-houses have a ventilation hole *(hava-kesh)* of about half a metre in diameter in the crown of the dome; the wind blowing across the dome would suck air out. Unfortunately the ruined state of the crown of the domes makes it impossible to say just how they were protected from the weather; presumably before a rainstorm sacking would be wrapped round the sides of the hava-kesh to keep the water out. Whether or not a hava-kesh is provided, constant evaporation occurs through the permeable mud-brick shell of the dome thus reducing both humidity and temperature. In thatched ice-houses (see below) constant ventilation could be simply achieved.

Ice making

Obtaining the ice was a much more difficult matter in Persia than in Britain where it was skimmed from a handy lake or pool and packed away early on a cold morning – a job traditionally organised by the head-gardener. On the Plateau water itself is not only in short supply, it is often saline; even a low degree of salinity would reduce its freezing temperature so the siting of ice-houses was limited by the availability of a pure water supply, a qanat being a likely source (see Chapter III).

Extremes of temperature add to the difficulties. In winter, in the desert, even when the temperature drops to freezing at sunset, the sun can be hot by mid-morning, which further reduces the chance of collecting enough ice to fill the vast pits. To make ice, a spot was chosen to which water could be diverted whenever it was needed; here a long shallow pool was dug out, about 50 x 10 metres and 40 to 50 cms (2 feet) deep. The excavated earth was used to build a high wall along the south side of the pool which shaded it from the sun and protected it from wind (figs 54 and 55); the width of the pool was governed by the protection given to it by the wall. The pool was lined with baked brick tiles to make it water-tight and to stop the water becoming muddy, and perhaps to insulate it from the ground. C.J. Wills probably gives the best first-hand description of ice-making: "The delicious Ab-i-Rookhni ("stream of Rookhnabad") is diverted from its course during the first cold night. A few inches of still water is collected in the pool, by morning it is frozen; at night the water is again admitted, and another inch or two of ice made. When three to six inches thick, the ice is broken and collected for storage in a deep well (ice pit) on the spot: and so day by day the process goes on during the short winter until the storehouses are full" (Wills 1891, 241).

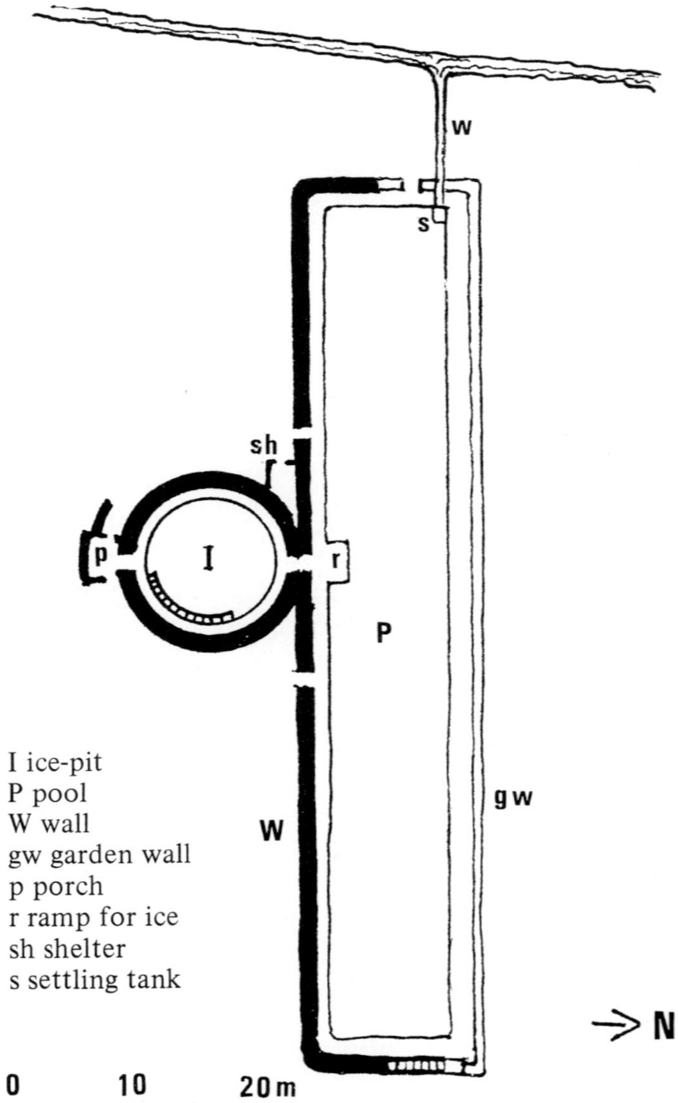

55. Plan of the ice-house near Yazd (drawn to a smaller scale than the section (fig 54).

Persians who remembered the work in the ice-houses told how on clear frosty nights the water would be diverted to the ice-pool and the ice formed was skimmed off in layers of 2 or 3 cms (c. 1 inch) thick. The thickness must have varied depending on the cold, and methods of collection seem to have differed from place to place.

Chardin (in Persia, 1762-77) described how in Isfahan a deep hole (an open ice pit) was dug and "before it they dig deep squares of sixteen or twenty inches, like so many little Basons; they fill 'em with Water over Night, when it begins to freeze, and in the Morning, when all is frozen, they break it to Bits, and put all these Pieces together in the Hole (the ice-pit), where they break 'em again into little Bits, as well as they can; for the more Ice is broken, the better it is; ...." At night more ice was made in the "basons" and the broken-up pieces of ice in the big ice-pit were themselves watered " .... To the End that they may hold the better together. In less than eight Days Working after this Manner, they have Pieces of ice five or six Foot thick: and then they gather the people of the Quarter together, who with loud Shouts of Joy and Fire lighted upon the Edges of the Hole, and with the Sound of Instruments to Animate them, go down into it, and lay these Lumps of Ice one upon the other, which they call *Codrouc*, as much as to say, Basis or Foundation, and throw Water between, to make them hold together the better."

Great care was taken to sweep away any snow that might happen to fall but, all being well, in "six Weeks time, an Ice-house of a Fathom or more deep, and as broad and long as one will, is fill'd to the very top with Ice." (Chardin 1927, 239). The ice, once rammed down in the pit, was insulated by layers of reed or straw, or "a sort of Sea-Rush, which they call Bazour" and then the door was sealed and the ice kept safe until needed in the summer. This system does not seem to have altered much over the years.

Chardin's observations of the cheerful Isfahan customs continue "In the Summer when they go to open the Ice-House, it is another Festival for that Quarter ..... What is very Remarkable, as well as Agreeable in their Ice, is its beauty and clearness ....". This is more than ca be claimed of ice skimmed from some lakes in Britain in the early years of this century.

At first sight the only function of the wall would seem to be to protect the pool from the sun. Walls are built along the long south side, with short lengths continuing east and west (fig. 55), so in winter, when the sun is lower in the sky, they completely shade the pool, thus minimising the warming of the tiled bottom and the ground under an empty pool or the thawing of ice in one that is in use. This allows for the collection of ice after sunrise which saves the difficulties of working in the dark. (Working on the Plateau, even early in the winter - late November and early December – made one realise how quickly hard-frozen ground thawed out. A plane-table, whose tripod legs were properly set up on hard ground at the start of a day's work, might unexpectedly sink and tilt only a couple of hours later, when the ground had thawed and softened).

The second function of the wall is to protect the pool from the wind. Moving water, even if only ruffled by a breeze, freezes more slowly than still water, and temperatures will only fall markedly when the air is still. Then each layer of cooled air near the surface stays where it is, because it is more dense, and so takes another turn of cooling, giving very low temperatures near the ground, perhaps several degrees lower than that of the air only a metre or so above it.

The importance of protecting the pool from the wind was stressed by the diplomat, James Morier, on an excursion made from Shiraz: "A wall is built the whole length of the reservoir to screen the ice from the south wind which is hottest" (Morier 1812, 123). No mention is made of shade from the sun. The French architect, Maxime Siroux, who considered ice-houses in his great work on caravanserais, went further "Les hautes murailles arrêtent le vent, ce qui provoque une prise rapide et régulière de la glace; elles permettent également – lors des grands froids – de confectionner des glaçons pendant la journée" (Siroux 1949, 131). He gives a variation between the shade and the sun temperatures of the order of fifteen to twenty degrees Centigrade (c.58° –67° F).

The line taken by the walls varies considerably. The straight ones already described seem to have been typical but others are curved; perhaps the most fascinating seen consisted of two curved walls and two ice-pits, like some strange winged creature (fig 58).

56. Curved shade walls. An ice-house near Sirjan. (1975).

57. Straight shade walls. An ice-house near Abarquh. The land rover gives an impression of scale (1970).

Siroux also cites instances (Kashan and Varamin) of places where the cold is short and sharp, and ice pools, unprotected by walls, were spread over several hectares. The ice was stored in domed pits (like those described above) which he seems to regard as unusual. The typical storage pit to him was a long narrow chamber, protected by a series of small domes, running along the base of the wall on its south side, the north being used for the ice pool.

Thévenot, who was in Isfahan at about the same time as Chardin, gives a similar description of Ice-making adding "This is an easie invention at Isfahan where the Air is very dry, and where there is little moist Weather.." (Lovell 1686, 96). The ice in Thévenot's description was stored in a long trench along the north side of the wall, "three fathom deep" and "and three to four feet" wide. This does not seem to have been roofed with mud-brick, but,when the ice in it was "a fathom and a half high", it was covered by a two to three foot thickness of straw and reeds (and perhaps it was also thatched to throw off the chance rain). Thatched pits or low buildings such as these could quickly be submerged in sand when they were no longer used. This might account for some of the huge free-standing walls with no obvious function which were still to be seen outside Tehran (near the Isfahan road) in 1961. They may have been ice walls.

Superficially, the idea of making ice in small "Basons" dug in the ground resembles the methods practised by the East India Company in the 18th and 19th century, and perhaps earlier, but there was a basic difference of principle between the ice-making on the Plateau, described above, and that of the East India Company. The latter was based on the loss of heat which occurs when water evaporates. It was described by Mrs. Fanny Parkes in *Wanderings of a Pilgrim in Search of the Picturesque*, her account of life in India during the 1820's. Her observations are not confined to the picturesque and are based on experience of a good many years. "Jan. 22nd - My husband has the management of the ice concern this year. It is now in full work, the weather bitterly cold, and we are making ice by evaporation almost every night. I may here remark, the work continued until the 19th of February, when the pit was closed with 3000 *mann* [over 107 tons] .... The ground belonging to the ice concern is divided into *keearees*, or shallow beds, very like saltern-pans in England, about six feet square and a cubit in depth; between them are raised paths. When the weather in December is cold enough to induce us to suppose water will freeze at night with artificial aid, the business of ice-making commences. At the bottom of the *keearees*, the shallow square beds, a black-looking straw is spread about a foot in depth, called "*pooal*," which is reckoned better for the purpose than wheat-straw ...." The ice was made in small unglazed earthenware pans which were filled from the pitchers seen in her illustration (fig 59). "The *pooal* straw in the shallow beds must be kept perfectly dry, to produce evaporation and the freezing of the water in the little pans placed upon it; should rain fall, the straw must be taken up and thoroughly dried before it can again be used." Evaporation would lower the temperature of the water and the dry straw would insulate

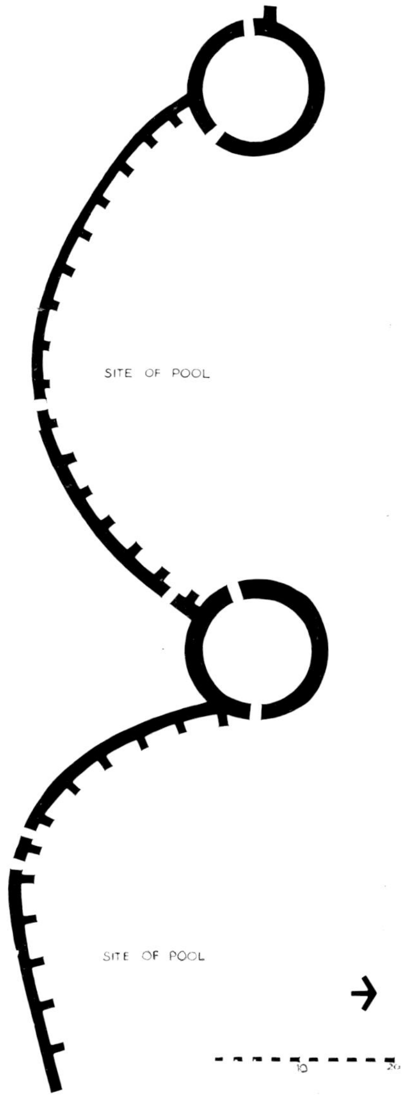

58. A pair of ruined ice-houses with curved shade walls near Sirjan. (1975).

9. Ice-making near Allaha-ad, c. 1828 as illustrated y Mrs. Fanny Parkes.

the pans from the heat radiated from the ground, thus the temperature of the water in the pans could be several degrees lower than that of the air above the surrounding ground. Indeed Mrs. Parkes claims that "The highest temperature at which ice was made in 1846, at Cawnpore, was 43° of Fahrenheit or 11° above freezing point". This is surprising but the thermometer may have been kept a few feet above the ice-pans and this, together with the loss of heat through evaporation, could account for the temperature difference. Incidentally, Mrs. Parkes' claim that "If the night be frosty, without wind, the ice will form perhaps an inch and a half in thickness in the pans. If a breeze should blow, it will often prevent the freezing of the water, except in those parts of the grounds that are sheltered from wind" confirms the opinion that a function of the shade wall is to give protection from the wind. The ice was stored in the thatched buildings which also appear in her picture. It was nice to read on the following page, that when poor Mr Parkes was struck by a fearful fever, and the surgeons could not decide whether to employ the lancet or claret, they opted for the latter – a bottle a day for three weeks. '... his head was enveloped in three bladders of ice and iced towels were around his neck.' He recovered. (Parkes I 1852, 78)

This Indian ice-making system had no advantage on the Plateau where the winters were much colder but it may well have been used elsewhere in Persia at lower altitudes where conditions were similar to those in Allahabad It is unlikely that the Persians were unaware of it. Ice was procurable as a matter of course at the best of the caravanserais. It was essential to slake one's thirst after the heat and grit of the journey before embarking on the exchange of news, views and ideas which are the currency of all travellers. The availability (or absence) of ice and probably its source must have featured in many a discussion.

An Englishman who went on a journey from Tehran to Kashan in May 1851 wrote:"The luxury of ice was to be had at nearly every stage; the ice receptacles having been opened at the approach of warm weather; and for a single shahee, one could procure as much of this as he required". (Binning 1857, 308).

60. A ruined ice-house near Taft, 1970.

5 WIND-CATCHERS

*Contributed by Susan Roaf**

In parts of the Plateau, particularly in the regions of Yazd and Kerman, the roofs of the settlements are dominated by mudbrick towers which look rather like chimneys with an assortment of vertical vents at their heads. These are the wind-catchers which contribute so much to the comfort of those living through intensely hot summers, and to the extraordinary skyline of those places where they abound. They were well described by John Fryer who wrote of the wind-catchers in Bandar Abbas in the late seventeenth century: "The structures are all plain atop, only *Ventoso*'s, or Funnels, for to let in the Air, the only thing requisite to living in this fiery Furnace with any comfort; wherefore no House is left without this contrivance; which shews gracefully at a distance on Board Ship, and makes the Town appear delightful enough to Beholders, giving at once a pleasing Spectacle to Strangers, and kind Refreshment to the inhabitants; for they are not only elegantly Adorned without, but conveniently Adapted for every Apartment to receive the cool Wind within " (Fryer II, 159f).

** I am grateful to Michael Harverson and Elisabeth Beazley for extensively revising and editing this chapter – S. R.*

61. A village in the desert, Chupanagh, 1975.

Badgir is a term used throughout Iran and the Gulf for these structures and it means literally a wind-catcher. They are most commonly found where winds can be relied on to blow at regular intervals, notably in the ports along the Persian Gulf and the settlements to the south-west of the central deserts. In Iran a wind-catcher consists of a tower which is usually subdivided by brick partitions to contain several shafts. At the head are vents in which the prevailing winds are caught and channelled down through these shafts. The rooms below are thus both air-conditioned and ventilated.

When there is little or no wind there is no noticeable movement of air down the shaft. Due to the heating up of the bricks of the tower during the day air in the shaft warms up and the shaft acts as an air stack (i.e. like a chimney), causing the hotter air to be expelled from the vents of the tower; the cooler air from the rooms below is then drawn into the tower and replaced in turn by air from the courtyard. This effect may occur at any time during the day, but is most common in the still air conditions at night.

When there is a wind, the positive pressure on the windward side of the tower is coupled with a negative pressure on the leeward side, where a vortex is created by the mass of the tower. When air travelling down the windward shafts reaches the base of the partitions inside some of it will leak up the leeward shafts of the tower, attracted by the negative pressure of the vortex. In stronger wind conditions this leakage is not important because the thrust of the downward moving airstream is out and into the room, but in conditions where air is moving out of the base of the tower at less than 0.3m (1ft)/sec , the leakage must represent a considerable proportion of the air in circulation.

It will be appreciated that the air introduced by a wind-catcher may be warmer than that of the room which it ventilates (see P. 66). This may seem to be a contradiction in terms of comfort, but the precise function of a wind-catcher is more easily understood when the basic need of a human body is considered. In hot climates heat must be lost from the body. Heat losses from the body by radiation and convection, the normal methods in a temperate climate, become progressively less effective as the ambient temperatures rise. Unless the body can sweat enough these methods cease altogether when the skin is at the same temperature as that of the surrounding air and of the walls, floor and ceiling of a room. Then the only way the body can lose heat is by evaporation, hence the importance of a draught. The drier the draught of air, the more effective it will be in evaporating sweat from the skin, thus cooling the body and giving a sense of well-being. It does not matter if the draught is slightly warmer than the ambient air: it will evaporate sweat and cool the body just the same.

History and distribution on the Plateau

62. House with two wind-catchers drawn on papyrus, found in the tombs of the New Kingdom in Egypt around 1500 B.C.

It is unfortunate that the first part of a building to collapse or be replaced is usually the roof and while archaeologists can tell us much about the history of ground plans it is far more difficult to ascertain the history of roof forms. Simple wind-catchers, however, were a very early invention and were depicted on the roofs of houses in pictographs on papyri found in the tombs of the New Kingdom in Egypt (1580-322 B.C.) (Badawy 1948, 86f). These pictures of wind-catchers (fig 62), known in Egypt as *malkaf*, bear a remarkable resemblance to those that can be seen on the roofs of Cairo today, but that is hardly surprising, as they represent a simple and effective technology. The basic nature of their design is illustrated by the inclusion of similar devices in the woven tent clot of the Cyrenaican Bedou. The wind-catcher, supported by three wooden poles and held in place by a guy-rope, is sat beneath during the day and provides effective air-conditioning (fig 63).

Thus we can envisage an early and fairly widespread simple model of wind-catcher, but the history of the mor elaborate examples is harder to trace. It has been suggest ed that wind-catchers were associated with the two recesses at the back of the great throne hall at Babylon (c.600 B.C.) and again with the sophisticated air circulation system at the Sassanian fire temple at Firuzabad (c.250 A.D.), but without more evidence it is impossible to justify these claims.

The first authenticated instances of wind-catchers w have on the Plateau are six 14th-century examples: one in Khorasan and five in Yazd province (O'Kane 1976, 85 These include two and four directional examples and sophisticated vent details. The Timurid period in Iran wa

63. The side of a bedou tent, showing the flap in the tent cloth which acts as a wind catcher.

an era of extensive building and a time when Yazd was an important centre of trade and intellectual activity. Many of the more elaborate vent details that can be seen on wind-catchers in Yazd today are careful copies of details that ornament Timurid buildings and there is a strong possibility that the most typical design of tall wind-catcher evolved during the same period.

17th-century travellers described wind-catchers in Isfahan, Shiraz, Lar, Kashan and Tabriz and it appears that they were widely used by this time. Thévenot referred to them in the following fashion: "At Shiras, Lar, and in other hot Countries, they have upon the tops of their houses, an invention for catching the fresh air: It is a Wall one or two fathome high, and bout the same breadth, to which at the intervals of about three foot, other Walls about three foot broad and as high as the great Wall, join in right Angles; there are several of such on each side of the great Wall, and all together support a Roof that covers them: The effect of this is, that whatsoever corner the Wind blows, it is straight betwixt three Walls, and the Roof overhead, and so easily descends into the house below, by a hole that is made for it." (Lovell, 1627, 87). This involved description which suggests surprisingly wide vents and stresses the roof, does not call to mind the elegant construction of Persian wind-catchers. In any case wind-catchers are more satisfactorily summarised in sketches or photos than in words.

The influence of the diligent Persian merchants can be seen on skylines all along the Gulf: as the trade routes became established and well-used in the 17th and 18th centuries between Yazd, Kerman and the Gulf the settlements of Lar, Bandar Abbas and Bushire evolved their

64. Windscoops on the skyline of Hyderabad, Sind (after Koenigsberger).

own styles of wind-catcher, taking advantage of the sea breezes. Trading between ports on both sides of the Gulf was heavy and as the Persian merchants settled in ports on the Arabian mainland they took with them the technology of the wind-catcher to ports such as Manama (on Bahrain), Dubai and thence perhaps as far as Hyderabad and Tatta in the province of Sind in modern Pakistan, towns noted for their distinctive simple wind-catchers (fig 64, based on Koenigsberger 1973, 226; also Coles and Jackson 1975, 2).

Despite the introduction at the beginning of this century of *pankeh* (ceiling fans) and, later, mechanical ventilation systems, wind-catchers are still considered cheap and efficient enough to be incorporated in many new buildings in the villages and cities of the Iranian Plateau (fig 61).

It was on the south-western fringes of the central deserts, in the towns of Kashan, Nain, Yazd and Kerman, that the most elaborate wind-catchers were built. Climatic conditions in this region are ideal for their use: intensely hot summers are moderated by the constant afternoon breezes, predominantly from the north and the north-west. Data from the Iranian meteorological office indicates that prevailing afternoon winds in Yazd are from the north-west and average 9m (29ft) per sec , while those in Kerman are from the north and average 10-14m (32-45 ft) per sec , thus creating good conditions for the effective functioning of the wind-catchers. Without a doubt Yazd, known as *shahr-i badgirha*, (city of wind-catchers) was the centre of evolution of the tall wind-catcher. Evidence can be seen today in the unusual skyline of the city where every stage in their development can be seen, with examples varying in age from those built last year to those that are six hundred years old.

In the province of Yazd almost every village has wind-catchers protruding above its roofs and in Yazd itself there are around a thousand wind-catchers: several are built on mosques, generally above the *mihrab*, ventilating and cooling the main prayer hall; two are built on caravanserais, over the main apartments on the western side; some forty are built beside water cisterns; while over ninety per cent are built on houses, where they are used to cool summer living rooms and to ventilate basements.

Types of wind-catcher

Wind-catchers can be divided into two main groups (fig 65): those built over ordinary houses and cisterns and the much grander ones built by rich householders and by benefactors of public buildings.

The plans of most ordinary wind-catchers may be relatively elaborate, but details of the head and base of their vents are at times simple, even crude. They usually rise less than 3m (c.10 ft) above the roof and in many cases no higher than 1.2m (c.4 ft). These are commonly built on the poorer houses. They are also used in larger establishments as secondary wind-catchers, as on the *haram serai* (harem) at the Bagh-i Dowlatabad in Yazd, where the main wind-catcher above the summer rooms is supplemented by a pair of smaller, simple wind-catchers above the corridors.

When a wealthy householder requires a prestigious wind-catcher he calls in a master builder to construct a large, imposing and elegantly detailed tower. These structures are between 2 and 22m (6 and 71ft) higher than the roofs on which they are built; most are 3-5m (10 -16ft), but in Yazd 110 wind-catchers rise over 5m (16ft). Their details are generally ornate and skilfully executed. Some have projecting heads which are designed to increase the volume of air that flows down the shafts of the tower. Many such details have been used by successive generations of builders for centuries. A builder will have a very fixed idea of what constitutes a good detail and a jerry builder will be ostracised for the crudeness of his details. In the last twenty years the traditional plaster vent details, with names like Turk's Head and Hubble-bubble (*qaliani*), have given way to standard brick versions which are given names such as 'the engineer's detail' (fig 68).

Design Details

The performance of a wind-catcher is affected by variables such as height, cross-sectional plan, the orientation of the tower and the location of its outlets.

Wind speeds increase with height above ground, due to decreased wind resistance from the ground surface, and air temperature decreases with height above the warm ground. Thus the height of the wind-catcher influences the speed and temperature of the wind captured in the vents. The overall height of the vents at the head of the tower will also have a bearing on the volume of air taken into the shaft. There are structural limitations on the heights of the vent sections of the towers, since their internal partitions are constructed with thin bricks standing on their sides, supported on cross timbers (fig 66). Many of the smaller wind-catchers, rising a mere 80cm (c.30 in) above the roof, could more realistically be described as ventilators, while the highest examples, rising up to 22m (c.72 ft) above the roof, induce considerable pressure gradients from top to bottom of the shafts simply by their height.

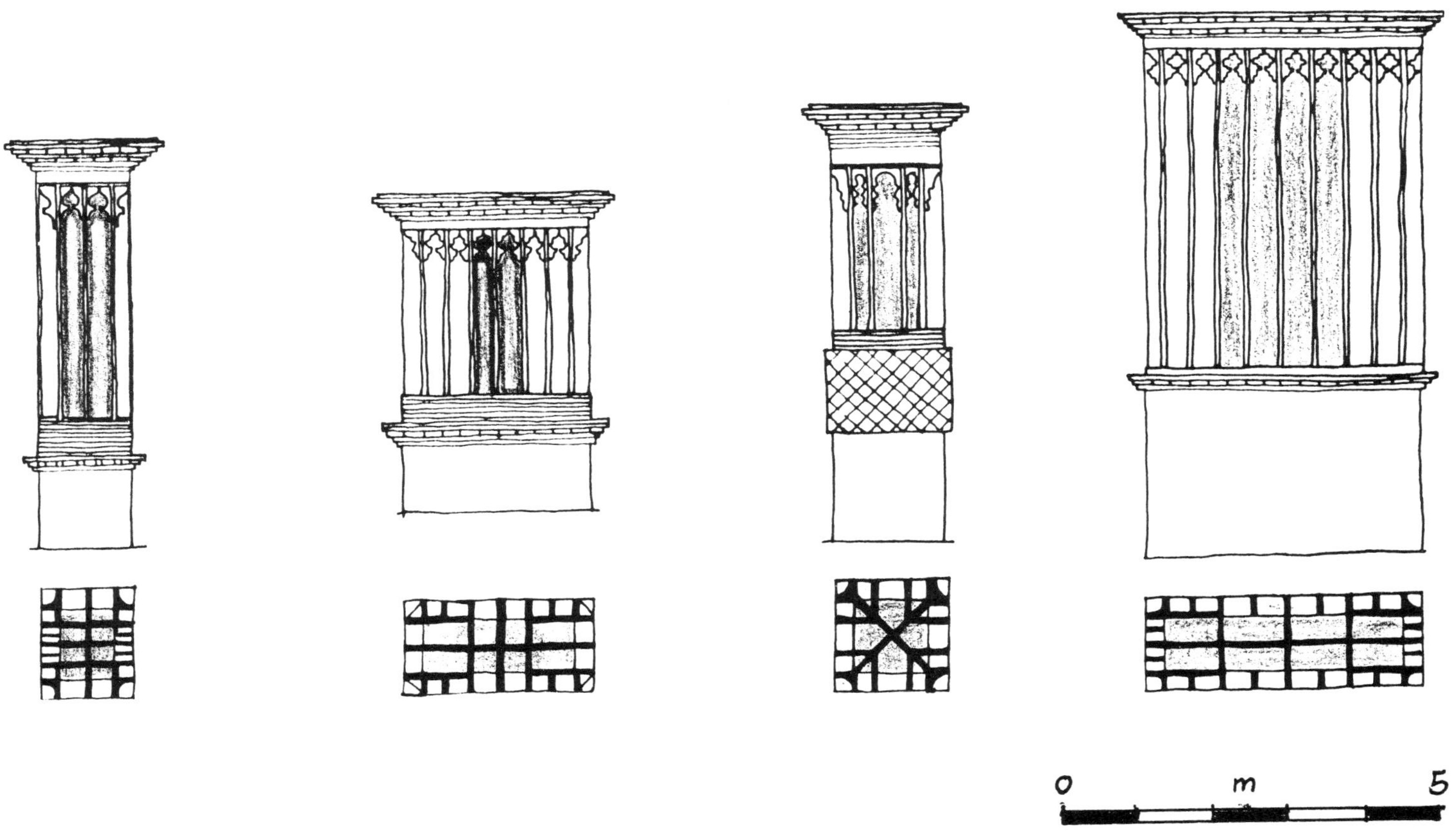

65. Plans and elevations of simple and elaborate wind-catchers.

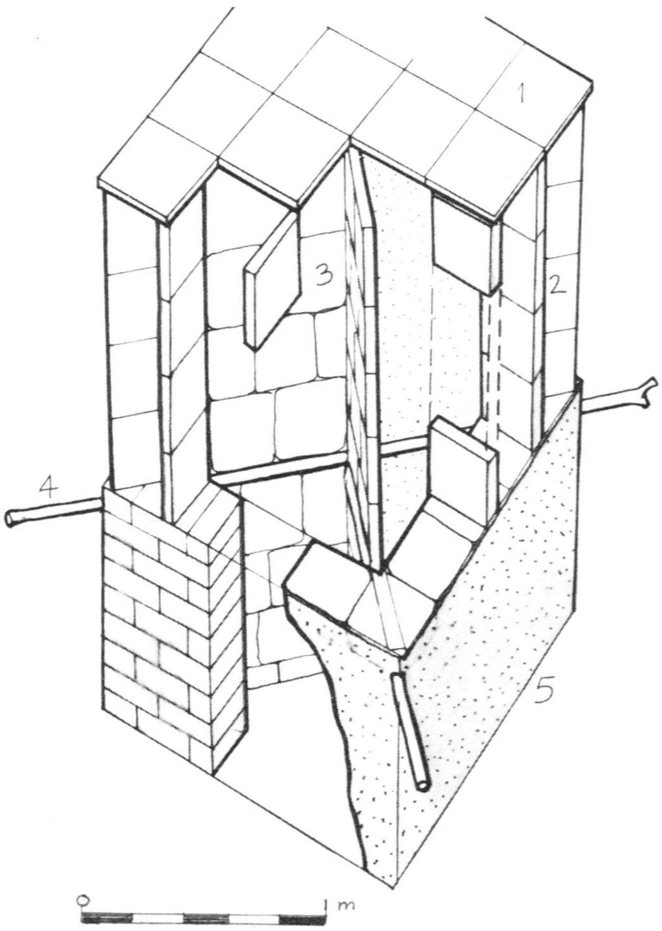

66. Axonometric drawings to show the construction of a wind-catcher.

The cross sectional areas of shafts determine the total volume of air that can pass down them and influence the speeds at which it travels. There is a wide variety in their size, ranging from 40 x 80 cm (c. 16 x 42 in) to 7 x 11m (c. 23 x 36 ft). The relationship between the proportions of the cross sectional areas of the shaft to the area of the vent openings is critical, for there comes a stage at which, by increasing the size of the shaft, air travelling down it is slowed to such a degree that the structure becomes inefficient.

Fig 67 shows that the plans of wind-catcher towers vary considerably. They may have one, two, three, four, five, six or eight sides open to the wind. Each side of the tower may itself have up to twelve separate vent openings which often remain distinct right down within the tower. The vents trap and channel air into the shaft and decrease turbulence in the tower by regulating the direction of the incoming air. The more sides to the tower the greater its ability to respond to changes in wind direction, but the smaller its intake of air from directions of predominant prevailing winds. Square towers are most common, closely followed by towers with sides in the ratio of 1:2, then 3:2, 5:2 and 3:1, with other combinations of side lengths following some way behind. Sides with the largest vent area usually face into the direction of the prevailing wind.

Common locations for outlets occur as shown in fig 68. From observations in Kuwait, Lewcock considers a

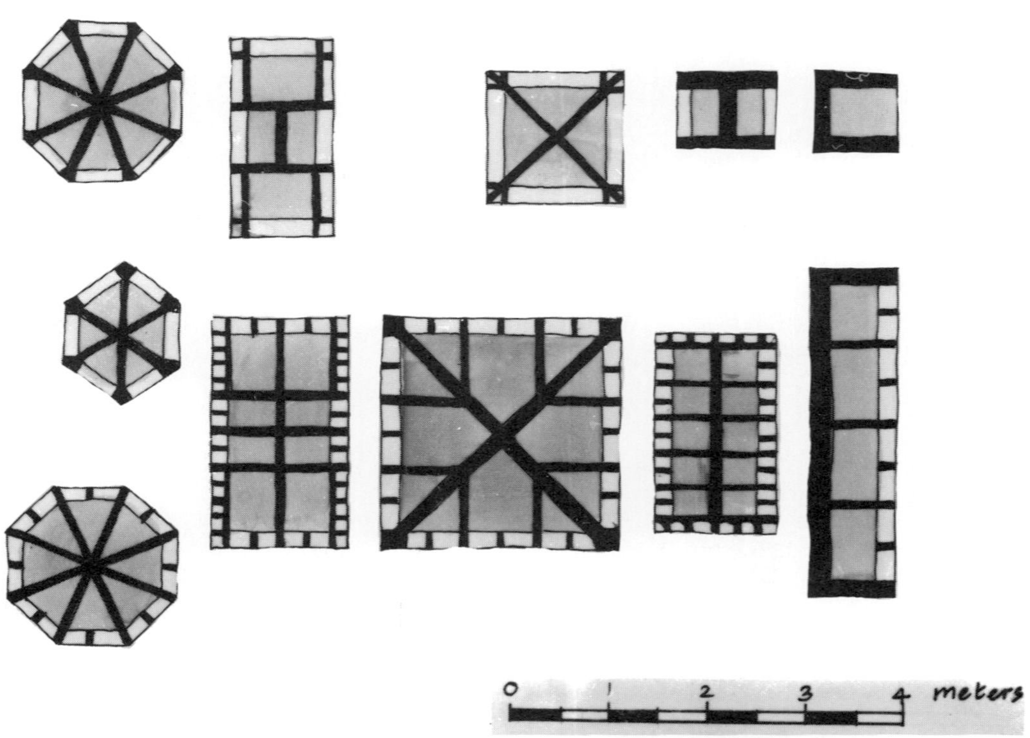

67. Plans of the various types of wind tower.

height no greater than 1.5 m (5ft) above the floor is most effective; he also describes the wind-catchers of Bahrain placed as horizontal flues midway up an external wall, capitalising on the high pressure outside and creating a draught low down in an upper room where people are sitting on the floor (Lewcock 1978, 41 ff). The climatic conditions on the Gulf are of course different from those on the Plateau, where wind-catchers are seldom linked to rooms higher than those on the ground floor.

It is difficult to estimate the importance of habit or fashion in the building of wind-catchers, but provincial builders usually talk with authority about one type of wind-catcher: that which is commonly built in their own villages. A well-designed and constructed wind-catcher is still a status symbol in the village. At present in the cities a house serviced with air-conditioning or mechanical ventilation carries more prestige, although attitudes are changing (see P. 72).

In the centre of Yazd the socio-economic organisation of the city at the turn of the century is well illustrated by the size and ornateness of the wind-catchers in the various quarters of the walled town. The ancient quarter around and to the east of the Masjid-i Jumeh has elaborate examples of medium size dating from the 17th and 18th centuries. The working-class quarters in the north and north-west of the town have low, one or two directional wind-catchers. Finally the area around the Chaqmaq mosque has the cream of the grand, multi-directional,

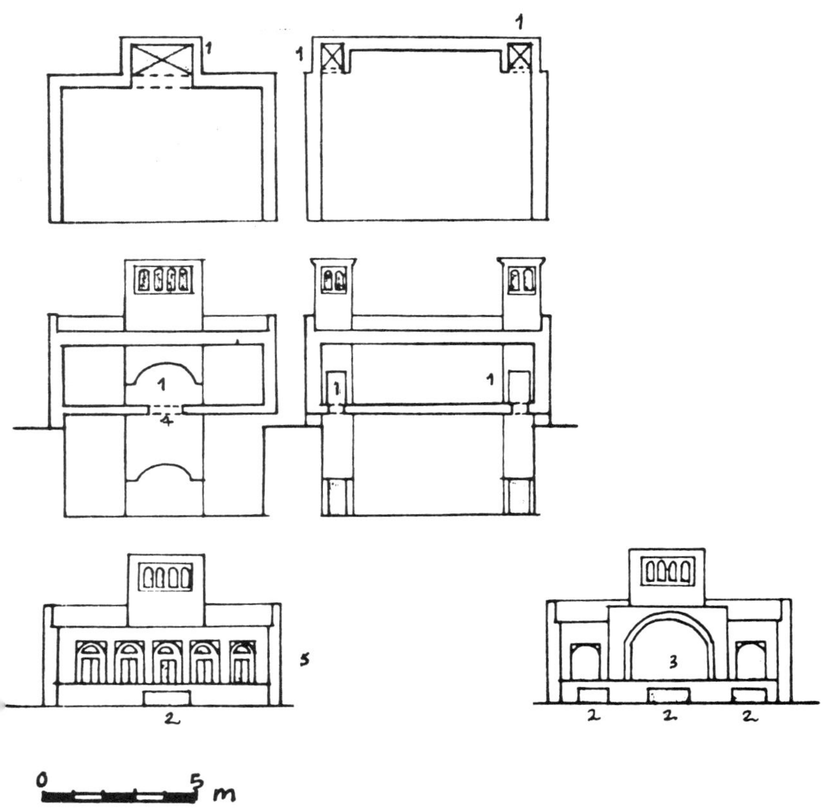

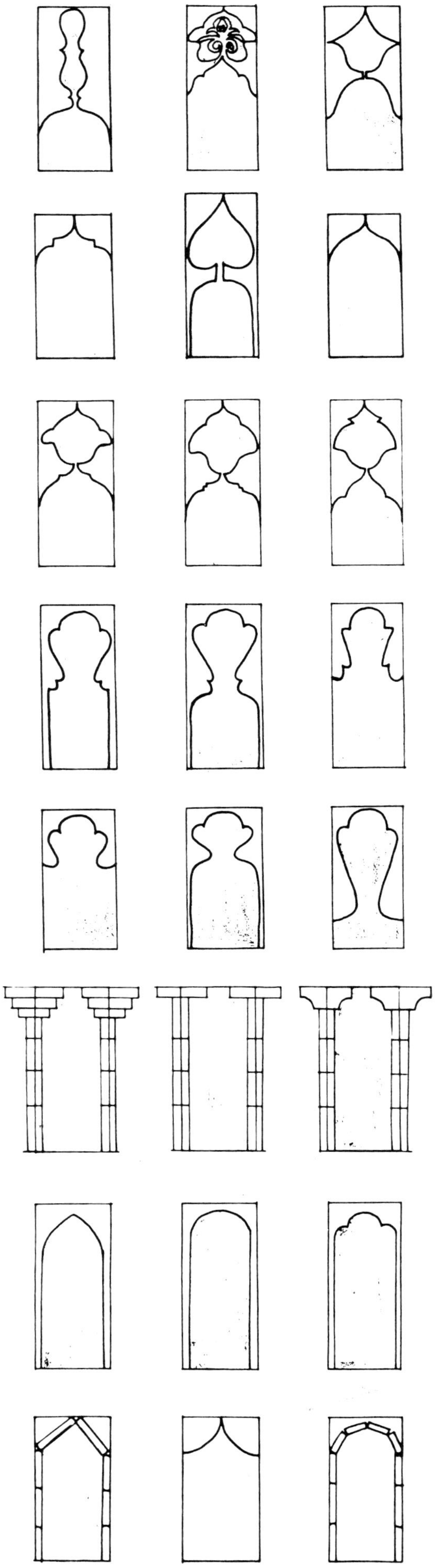

68. (above) Plans, sections and elevations which show common positions of wind-catchers. 1 wind-catcher outlets; 2 basement windows; 3 talar; 4 wooden grille; 5 five-door summer living room. (right) Details found at the heads of vents.

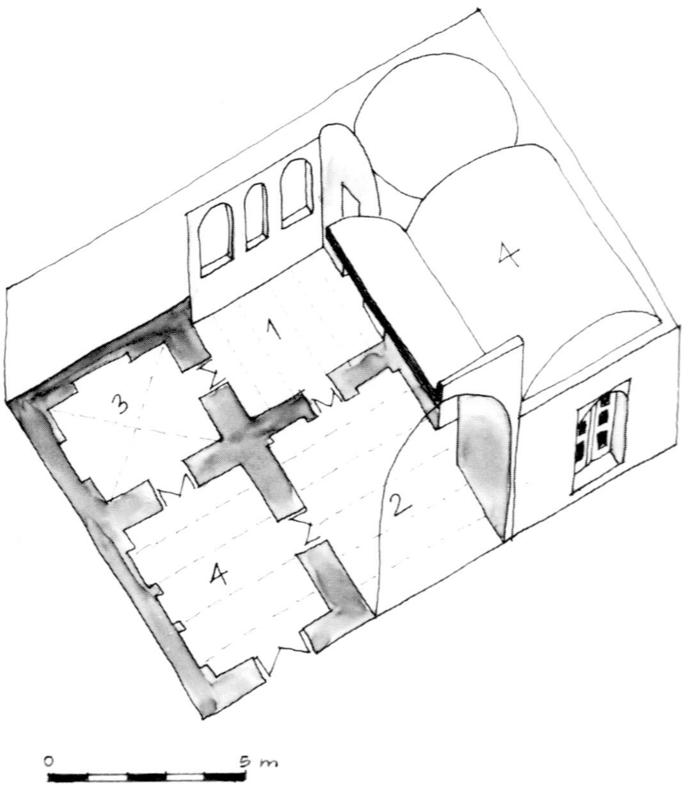

69. Axonometric drawing of a house at Aliabad without a wind-catcher. 1 summer living-room; 2 aivan; 3 store; 4 winter living-room.

domestic wind-catchers built during the 19th century. Zoroastrians, an important religious minority group, were forbidden until 1898 to build wind-catchers on their houses, but subsequently some of the finest were built by them in Yazd and in predominantly Zoroastrian villages such as Aliabad, Nausratabad and Zarch (north-west of Yazd, towards Ardekan).

Other more subtle influences are at play relating to the types and directions of the wind-catchers built. Where there is only a minimal buffer zone of cultivation between buildings and the desert care is taken to exclude the hot, turbulent winds from the south-east, so wind-catchers tend to face towards the north or north-west. Where, however, the buffer zone is large enough, dust and sand settle on the gardens and the air which passes over the vegetation is cooled and humidified in its passage and may be trapped in multi-directional wind-catchers.

The house beneath a wind-catcher

In order to understand the importance of the domestic wind-catcher one must first understand how the houses themselves are designed to respond to the extreme summer climate (see also Chapter II Pp. 26-28). This applies to most domestic buildings, from a simple village house to an elaborate garden pavilion.

70. A Yazdi house showing the various areas used during summer, viewed from the north. 1 roof; 2 talar; 3 hoz khaneh; 4 wind-catcher; 5 west facing wall blank; 6 family living-room; 7 guest courtyard; 8 wood/oil store; 9 well; 10 water storage tank; 11 kitchen; 12 stables; 13 qanat; 14 courtyard garden; 15 pool; 16 basement living area/store.

The house in Aliabad shown in fig 69 demonstrates how the climate may be manipulated for relative comfort even without a wind-catcher. The central room on the north side of the house into which the family retires during the heat of the day in summertime has its walls isolated from solar radiation to the south by the traditional Persian porch *(aivan)* and to the east and west by the store rooms. The north wall of the room is exposed, but it receives no solar radiation because it is shaded by the roof of the aivan. The two winter rooms on either side of the aivan have French doors facing south-west, through which the winter sun enters the rooms and heats the walls and floors, so turning the entire structure into a crude storage heater. The roof of the aivan is raised a metre above the general roof level of the house and faces a suitable prevailing wind. Air is caught in the barrel vault and is channelled into the back of and around the aivan. This increase in air speeds is sufficient to make the effective temperature almost comfortable.

The village house at Bunderabad in fig 71 exhibits a similar organisation of rooms, with the house facing south-east. However, the winter rooms here with their double French doors completely insulate the summer room from the penetration of sunlight and there is no roofed aivan: it is represented by a terrace raised 30 cms (1 ft) above the well-treed garden. A centrally placed pool is used for domestic purposes, for decoration and to cool and humidify the atmosphere. To the classic

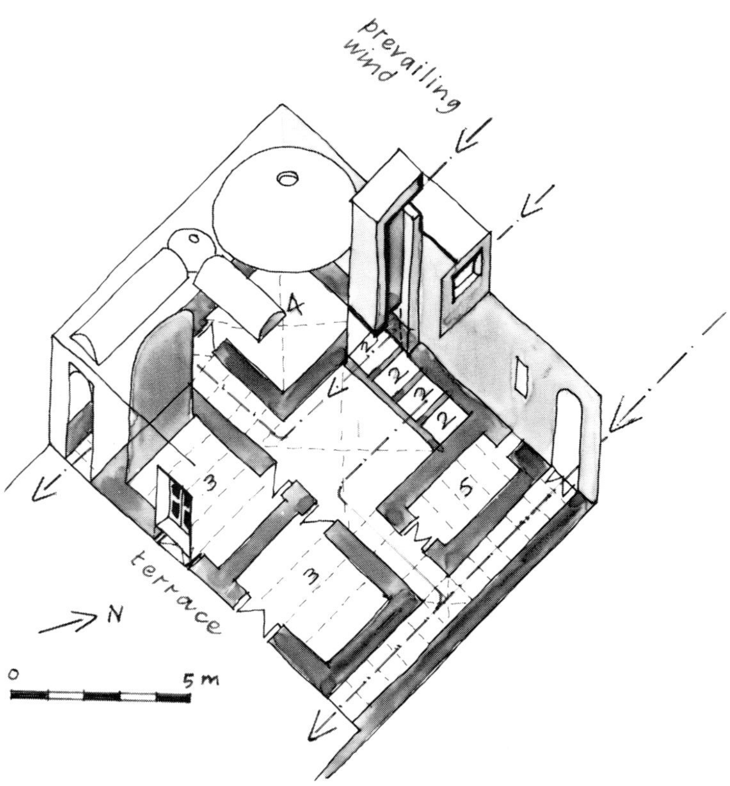

71. Village house with wind-catcher. 1 summer living-room; 2 wind-catcher; 3 winter living-room; 4 kitchen; 5 store.

combination of rooms is added a more sophisticated air circulation system which includes a wind-catcher, ventilators and corridors. At the north-west end of the dark, cool summer room is situated the wind catcher with its internal lintel 1.8 m (c. 6 ft) above the floor and its exterior vents facing north-west. The prevailing wind is channelled down under pressure through the vents and shaft of the wind-catcher into the summer living room and out of the open corridors on each side of the winter rooms into the area of low pressure on the leeward side of the building; the entrance passage also increases the through draught. Such corridors also convectively cool the external wall surfaces of the rooms they serve. They are used almost universally in this region to encourage air circulation through the houses because without the convective cooling effects of an air stream flowing at 0.5 - 1.5 m (1ft 8ins - 5ft) per sec. it is impossible in this climate to achieve, above ground, temperatures that fall within the comfort zone using passive cooling systems (any arrangement whereby there is no input of energy except that provided by the natural environment).

Traditional houses in Yazd itself are built around an enclosed courtyard, with the summer rooms of the house on the south side, facing away from the summer sun, and the winter rooms on the north side, facing towards the winter sun (fig 70). The rooms to the east and west of the courtyard are used throughout the year for a variety of purposes and the west-facing wall often has no rooms in it at all, as during the summer afternoons the high level of solar radiation on this wall, combined with the hottest ambient air temperatures, makes rooms on this side unbearable to live in.

The summer rooms on the ground floor, which is generally about 60 - 80 cm (c. 2 - 2 ft 6 ins) above the courtyard level, include either a main summer room with three, five or seven pairs of French doors onto the courtyard, which are kept open during the summer, or in its place a *talar*. This latter room, open on the north side, has a barrel-vaulted roof, raised up to 2 m. (6ft 6 ins) above the general roof level, which catches the prevailing winds so that they circulate within the space below and out into the courtyard. In the basement beneath the talar is a large summer living room, which may be flanked by subsidiary living and storage areas.

A single wind-catcher or a pair of wind-catchers may be situated at the rear of the summer living-rooms, with their shafts entering the ground-floor room through an opening generally 1.5 - 1.8 m (c. 5 - 6 ft) high or a door from the shaft of the wind-catcher. Directly below the centre of the shaft there may be a timber grille in the floor through which air flows into the basement, and on the courtyard side of the basement room are high level windows through which air is expelled or drawn into the room (fig 68). A local painting of c.1900 shows these details in a scene of Yazdis relaxing in a talar (Malcolm 1905, 26).

As the prevailing wind enters the tower through the vents at its head the air at high pressure piles up against the partition and is sucked down the shaft towards the lower pressure zone at the base of the shaft. As the air passes down the shaft it loses some of its heat to the walls of the shaft and may absorb moisture from the mud bricks of which the tower is constructed, moisture that has been absorbed from air passing down the shaft at night time. It has been generally thought that the passage of air down the shaft cools the air sufficiently to create an air-conditioning system, but measurements taken in Yazd indicate that the difference between the ambient shade temperature and the temperature of the air leaving the base of the wind-catcher shafts at the ground-floor level is only in the region of 1–2° C. (2–4° F.) in towers below 8m. (c. 26ft) high (fig 74). Often air introduced into the summer room may be at the same temperature as the room or even warmer, especially in the case of the basement, and the most important property of the air is that it is moving at a speed of up to 1m(c.3ft)/sec across the talar. Exceptional draughts attaining 3m (c.10ft)/sec are fairly regularly experienced across the summer rooms, causing from 10 to 30 air changes per hour. The physiological cooling, the heat loss from bodies due to evaporative cooling on the skin surface, may not be as important as the psychological effect of sitting in a stream of moving air.

In some Yazdi homes the family will sleep during the afternoon at the back of the talar near the wind-catcher outlet and in some cases right below the shaft, although in most houses the family will retire to the basement on summer afternoons. Basements in Yazd range in depth below ground from 3.5 to 5 m (c. 11 ft 6 ins to 16 ft 6 ins) and many houses have extensive complexes of store rooms and living rooms beneath them. Many basements do not have wind-catchers and, as would be expected, these have stable temperatures ranging from perhaps 25–28°C. (77–82°F.). Relative humidities (see Chapter III) in the middle of the day in the basement may be 40 - 65% and in the course of the night may rise as high as 75%. At high temperatures and in places where people are accustomed to very dry air such a rise in the level of relative humidity is unpleasant as the air becomes stale and dank. Basements with no wind-catchers are generally vacated at about 5 p.m. and the family moves up to the courtyard and later up to the roof.

Thus in basement rooms a wind-catcher has the function of introducing fresh air and getting rid of stale. Whereas an unventilated basement occupied by several people becomes uncomfortably humid, even in desert conditions, the air introduced by a wind-catcher is drier as well as fresher. The fact that it may be warmer is relatively unimportant.

The introduction of warmer air during the day causes fluctuations of temperature which are greater than in those without wind-catchers. In the well-constructed basement in the harem of the Bagh-i Dowlatabad (P. 60) summer midday temperatures are around 30°C. (86° F.); by late afternoon they have risen to c.31° C. (88° F.) and by midnight they are down to around 27.5° C (81°F.). Relative humidities, however, range only from 25% at 4 p.m. to 35% at 2 a.m., a variation which might scarcely be noticeable and is largely accounted for by the changes in temperature. So a basement with a wind-catcher may be slept in on hot summer evenings.

Living with a wind-catcher

The Yazdi is by nature a sedentary nomad: instead of the long distance migrations practised by the nomads of Iran he migrates through and around his house. Until about twenty years ago it was common practice for middle -class families in Yazd to move to summer quarters (*yeilaq*) about thirty-five miles south of Yazd in the Shir Kuh mountain range. The summer months would be spent in the fruit-producing valleys some thousand feet higher than the city. Here Yazdis had houses in gardens to which the whole family would move. Only the head of the household remained in the city, with a house-keeper, to continue his business, visiting his family only occasionally during the three month holiday. If, however, a family remained in Yazd throughout the summer without mechanical ventilation or in winter with little heating, the year might still be passed relatively comfortably by gently migrating around the house.

The orientation of a traditional Yazdi house is critical and with the organisation of the rooms around a courtyard a major migration is effected each year in September or October to the winter rooms on the south-facing side of the courtyard and in April or May across to the north-facing rooms. Exclusion or inclusion of the sun is the most effective form of climatic control used. Most of the winter months will be spent in the winter rooms on the ground-floor level, but in summer, due to the dramatic change between the night and daytime climates, comfortable conditions are achieved by a rather distinctive form of migration, not horizontally around the house as occurs from season to season, but vertically through the house.

Yazdi men, having slept on the roof, will rise once the sun begins to become uncomfortable and go to work from 7.30 a.m. until about 1 p.m. The hottest part of the day from 1 - 5 p.m. is usually spent resting in the basement. The women will have been occupied in household chores in the courtyard, talar or kitchen until lunchtime when they join the rest of the family for the largest meal of the day in the basement. Many men, especially those working in the bazaar, will recommence their occupations in the late afternoon, not returning home until 8 p.m. The evening is passed in the talar or on the aivan – here in Yazd an elevated terrace, usually about 2 - 3 m (c. 6 - 10 ft) wide on the south side of the courtyard. The family will retire late at night to sleep on the roof. The wind-catcher will have been sat, slept or eaten beneath: in the morning and evening in the talar, and during the afternoon and sometimes the night in the basement.

Fifteen to twenty people will commonly live in the larger houses of Yazd and in order to minimise the work for any individual in a family a strict division of labour is practised. The large distances that have to be travelled to complete simple chores in such houses – perhaps 30 metres (33 yards) to open a door or serve food from the kitchen to the basement or collect a pillow from the roof – would make living there an exhausting occupation for a single couple, but with large extended families children are employed to perform many of the everyday chores inside and outside the house. Shopping will be done by young boys because women in the better families traditionally never leave the house except to go to the mosque and the hammam, to visit other houses or to travel to the summer quarters. Cooking will be done by the mother or the older girls, while the younger children wait at table, open the front door and so on. The continuation of the extended family is essential for the survival of the large traditional house types in the area.

Thus their architecture, their character and use have for centuries been evolving with the society and its way of life to produce a successful solution to the problems of living with the desert. To change the nature of any one of these influences will inevitably disrupt the equilibrium of the whole solution. For instance, the introduction of mechanical cooling systems, which cannot be used in the large old courtyard houses, has led to changes in house types, social organisation and the way of life of the younger married couples in Yazd, changes which would not have occurred without the electricity that powers the coolers. It is difficult to evaluate the implications of this change, but the equilibrium has been disturbed and a new order must evolve. Whether the wind catcher will re-establish its important role in domestic architecture remains to be seen.

The biggest wind-catcher in Iran

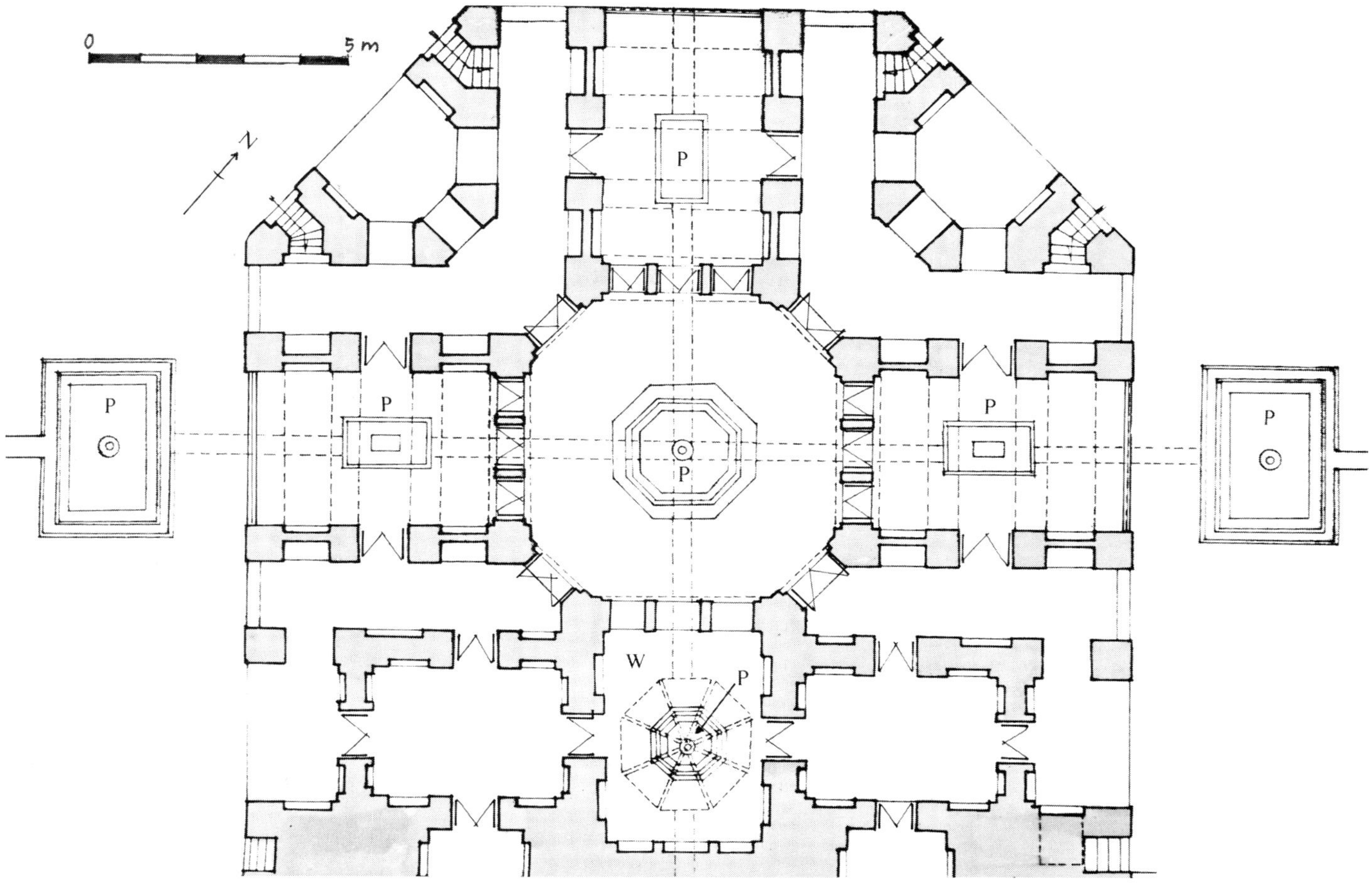

72. Plan of the great wind-catcher on the Khan's pavilion at Bagh-i Dowlatabad. P pool; W wind-catcher. The wind flows straight through the pavilion over the pools when the doors to the east and west of the central room are closed.

73. The great wind-catcher on the Khan's pavilion at Bagh-i Dowlatabad. (under repair 1976).

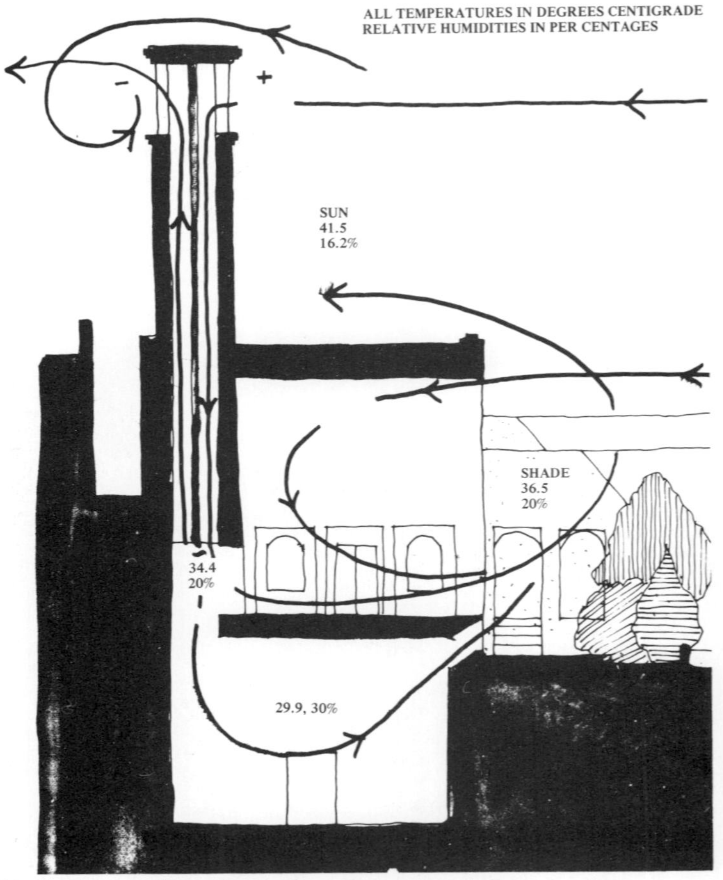

74. Section to show the performance of a wind-catcher in a Yazdi house.

Perhaps the most splendid wind-catcher ever built serves the Khan's pavilion at the Bagh-i Dowlatabad in Yazd. It rises 33.35 m (c. 108 ft) above the ground, with c. 9 m (30 ft) high vents at the head of its octagonal tower, built at the S.S.E. side of the pavilion (fig 73). The main axis of the garden runs from S.S.E to N.N.W. and the three main summer living rooms face S.S.W., N.N.W. and N.N.E., with an internal central room, the *hoz khaneh* (pool room), to the S.S.E. of which lies the wind-catcher room (fig 72). All the rooms are linked by a series of double doors which can be shut or opened to form an effective control system for the wind, which moves out of the wind-catcher room at velocities of 0.1 to 13 m (4 ins to 42 ft) per second. and average speeds of 3 - 7 m (10 - 23 ft) per sec. Each room has a pool in it, including the wind-catcher room, where at the base of the great tower, approximately 1.8 m (6 ft) below the bottom of the partitions that divide it internally, is a pool 1 m in diameter and 30 cm deep. Air passing down the tower has to travel over the surface of the water on its path into the rest of the building. In so doing it cools the water, as heat is lost through the process of evaporating water from the surface of the pool and the air is humidified and cooled at the same time.

The beauty of this building is that it can be used in so many different ways according to particular climatic requirements at any time of the day or year. In summertime the khan and his entourage would move around the building in order to avoid the intense solar radiation levels from the east and west during the day. In the morning he would breakfast in the west or north-facing rooms with a view of his harem or the length of the garden. In the heat of the afternoon he could shut the doors from the west and east-facing rooms and retire into the central pool room and sleep in the draught moving directly from the wind-catcher out through the doors to the north. In the evenings he could sit out in the east-facing room with the doors to the north and west closed, so enjoying the air flowing from the wind-catcher in a room whose walls had not received direct sunlight for over eight hours.

The combination of the strong draughts and the surrounding pools makes the khan's pavilion comfortable to live and work in without air-conditioning. The lesson this exceptionally well-designed building can teach is that with careful orientation, with the use of water internally and with thoughtfully planned air flow systems comfortable conditions can be maintained inside a house during the hot summer months.

The use of water with wind-catchers

Enough has now been said to refute the common misconception that wind-catchers cool the air and to suggest that pools directly beneath them do indeed achieve that aim. This latter solution, common in Egypt, is, perhaps surprisingly, rare in Iran, though pools are often included in the summer ground-floor rooms, sometimes taking up to one sixth of the floor space and being over one metre (3 ft) deep. In some of the oldest houses in Yazd courtyards were two storeys high, with rooms on the upper level used in the morning or evening and the ground-floor rooms used in the afternoon. There the pool room might be directly below the talar in the form of a semi-basement, with a large, open, arched doorway into the courtyard. The relationship between the wind-catcher and such pools can be seen in figs 70 and 72.

Air cooling may be provided in a true basement by associating this room and its wind-catcher with an underground water tunnel or qanat (fig 75). Under Yazd there exists a complex network of water channels fed by qanats from the mountains to the south of the city. Drawn down from the mountains with the water of the deep aquifers is cold air, which is used in some basements to cool and humidify the underground spaces of the house. The qanat may be linked to the basement by a vertical shaft 60 - 80 cm (c. 2 ft - 2 ft 6 ins) in diameter. Air at the head of this shaft has been measured to be some 6°C. (11°F.) cooler than the air in the other areas of the basement.

One of the most sophisticated examples of the use of a wind-catcher in association with a qanat is at the Bagh-i Khan, 5 km (3 miles) outside Yazd. Here the wind-catcher tower is situated 50 m (54 yards) away from the actual pavilion and is linked to its basement by an underground tunnel which crosses over and is connected by a perpendicular shaft to a deep qanat running from the mountains into Yazd. Air moving from the wind-catcher to the house mingles with the air from the qanat and is as a result cooler and more humid. The present owner has installed an electric motor in the passage behind the wind-catcher outlet in the basement, which will draw air through the system at a speed of up to 5 m (16 ft) per sec, though this refinement is not used on most afternoons.

This method of cooling basements for afternoon occupation precludes an Egyptian practice whereby a porous pot full of water is suspended in the shaft of the wind-catcher and beneath it is placed a grille covered with charcoal so that the air passing down the shaft is cooled, humidified and purified during its passage into the room. Another way of cooling the air in wind-catchers, which is worth mentioning since the wind-catchers there may well derive from Yazdi prototypes, is used in the Persian Gulf settlements, where large single directional wind-catchers are built facing the sea breezes. In their open vents are often placed loose screens made of straw, rushes or palm fronds, over which water is thrown during the middle of the day so that any air passing through the screen during the afternoon is cooled evaporatively and humidified in its progress into the summer rooms below (fig 76). This is not practised on the Plateau, perhaps because families generally retire to the basements which are relatively humid anyway in the afternoon.

75. Section showing how a qanat can cool a basement.
1 – 36.5°C, 20% RH; 2 – 29.9°C, 30% RH; 3 – 24°C, 45% RH; 4 – qanat.

76. A wind-catcher in Bandar Abbas with a palm frond mat over its vent which is soaked with water on summer afternoons. 1976.

77. *A water cistern with four wind-catchers near Fahraj. (1970).*

78. *Wind-catcher in the Kerman bazaar. (1977).*

Wind-catchers and water cisterns

Wind-catchers are integral to the construction of about half the cisterns in the Yazd and Kerman areas an of some elsewhere. Cisterns are discussed in Chapter III (Pp. 39f). Wind-catchers are commonly found in pairs flanking a cistern or in fours placed around it, although any number between one and six occur, facing in different directions (fig. 77). An interesting arrangement distinguishes one cistern near Nain where a wind-catcher o one side of the dome is a four directional square tower, while the opposite tower is square as well but has no internal partitions. Cisterns with wind-catchers are still built in the villages of these provinces (fig 45).

The future of the wind-catcher

Today many modern buildings are designed in a fashion that is ill-suited to the climate and can be considered habitable only as long as the electricity exists to power the mechanical ventilation systems. In Ahwaz in 1976 during a bad heatwave power failures were experienced due to the excessive load put on the electricity supply when all the mechanical coolers of the city were turned on full; fifteen people died from heat stroke during three days. Meanwhile, traditional houses remained comfortable inside. In the words of Hassan Fathy:

> "The architect who builds a solar furnace and then introduces a vast refrigerating plant to make it habitable is underestimating the complexity of the problem and working below the proper standards of architecture." (Fathy 1970. 6).

In villages all around the central deserts wind-catchers are still commonly built on new houses and provide the most important form of air-conditioning in many areas. These simple wind-catchers may cost no more than the time of the owner of the house and his friends to make mud bricks and construct the tower and perhaps the cost of wood used as structural timbers. They are cheap both in terms of initial outlay and running costs. Even a prestigious tower designed by a master builder will cost nothing to run. Only an earthquake is likely to cause it to fail. Each village that has wind-catchers will have its own particular style, and there are considerable regional variations. In Khorasan and Sistan, for example, there are villages where wind-catchers resemble small chimneys or 60 cm (c. 2 ft) high ship's ventilation ducts, placed on the apex of the domes above living rooms and facing, fully open on that one side, in one direction only (Michaud 1980, 21).

79. A wind-catcher on a Yazdi house. (1976).

80. Unidirectional wind-catcher being repaired on a village house in Chupanagh. (1975).

With the rapid urban expansion in Iran in the 1970's suburbs blossomed on the outskirts of old towns like Yazd. These included new types of wind-catchers, for instance on new bungalow estates where the houses have basements that are ventilated by modest baked brick wind-catchers with simple brick details at their heads. The wind-catchers are used because they provide effective ventilation for basements and because they enhance the skyline of the area and add a fashionable traditional element to the appearance of the houses.

It is encouraging to see wind-catchers also being used throughout the Middle East on large scale schemes such as the Museum of Modern Art in Tehran, on Qatar University, and on the New Baghdad conservation area scheme. Wind-catchers in these contexts have different functions: as light wells, ventilators and air-conditioners and for aesthetic reasons to articulate roofscapes.

The wind-catcher will probably be used for centuries in the traditional villages of the Plateau and it may also be used effectively and much more widely on modern buildings. Although some architects are now incorporating wind-catchers as air-conditioners and ventilators the device can only realise its potential as such when sensibly constructed and used in buildings which have been properly planned for the climate. It can then help to provide comfortable and pleasant living conditions without excessive reliance on expensive mechanical methods of ventilation and air-conditioning.

81. Sectional elevation of an idealised village from the central desert showing all the architectural features discussed in this book. (based on Khuranaq).

6 WATERMILLS

82. Watermill near the Doshmanziari Mamasani village north-west of Shiraz. A leat drawing water from a perennial stream terminates in the stone penstock directly above the mill (photo 1976, Susan Wright)

Fleeing across the Plateau before the Arab conquerors, Yazdigerd III, the last Sassanian monarch, was murdered in a mill in 651 A.D. at Merv, in the extreme north-east corner of his dominions. A century ago a British journalist, gathering information about this little-known area of Asia where Russia and Britain were engaged in 'the Great Game', observed at Merv "a very large number of turbine and overshot mills, by means of which the principal grinding of corn for the district is done" (O'Donovan 1882, II, 238). Such watermills still exist on the Plateau, though comparatively few work to capacity or indeed at all, except in remote or poor communities. The same is true deeper in the Alborz and Zagros mountains; some of those watermills receive mention here because study of them reveals an identity of design and operation and a similar role played in the local community until recently.

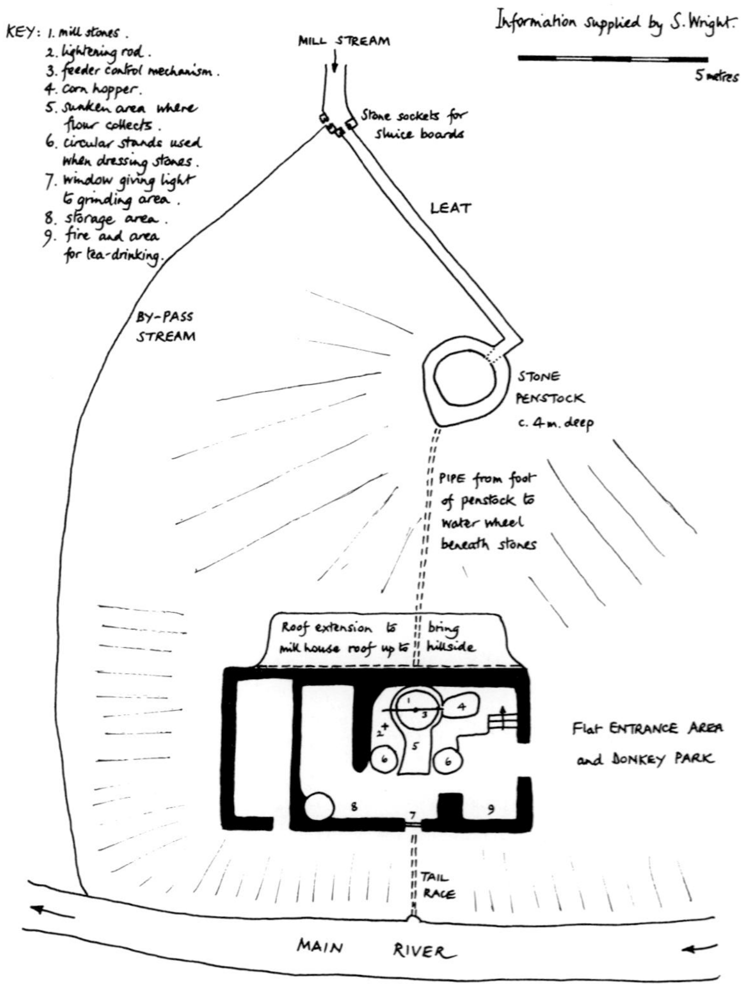

83. Plan of the Doshman Ziari mill.

Persian watermills generally involve no more than two pairs of stones, more frequently one. They are not large commercial undertakings, but cater for the needs of their local community. They are to be found under the landscape rather than dominating a stretch of river as so many fine English examples do.

Vitruvian mills occur on the Plateau, but most watermills there are of the type variously known as horizontal, Greek or Norse. Vitruvian ("overshot" in the quotation) mills are those with which we are most familiar in England: the water falls on paddles fixed to a horizontal drive-shaft which is geared within the millhouse to drive the stones (fig 84). In horizontal ("turbine" in the quotation) mills the paddles are attached to a vertical shaft which usually drives one pair of stones directly above it, without benefit of gearing (fig 85). These latter, known from China to the Shetlands, are of great antiquity in Iran. Their design enables man to harness a small water supply and to grind slowly, but steadily, with a minimum of supervision and skilled maintenance. By contrast Vitruvian mills require the steady, open flow of water which is rare on the Plateau.

Water to power the mills

Crucial to the efficient running of a horizontal mill is the provision of an adequate and well-regulated head of water. A disused example seen at Ghorian was turned by water from an irrigation channel flowing down a wide, gently sloping wooden shute barely 2 m (c. 6 ft) long. Simple watermills of this kind, with a daily capacity limited to about 160 kg (360 lb) were once common also

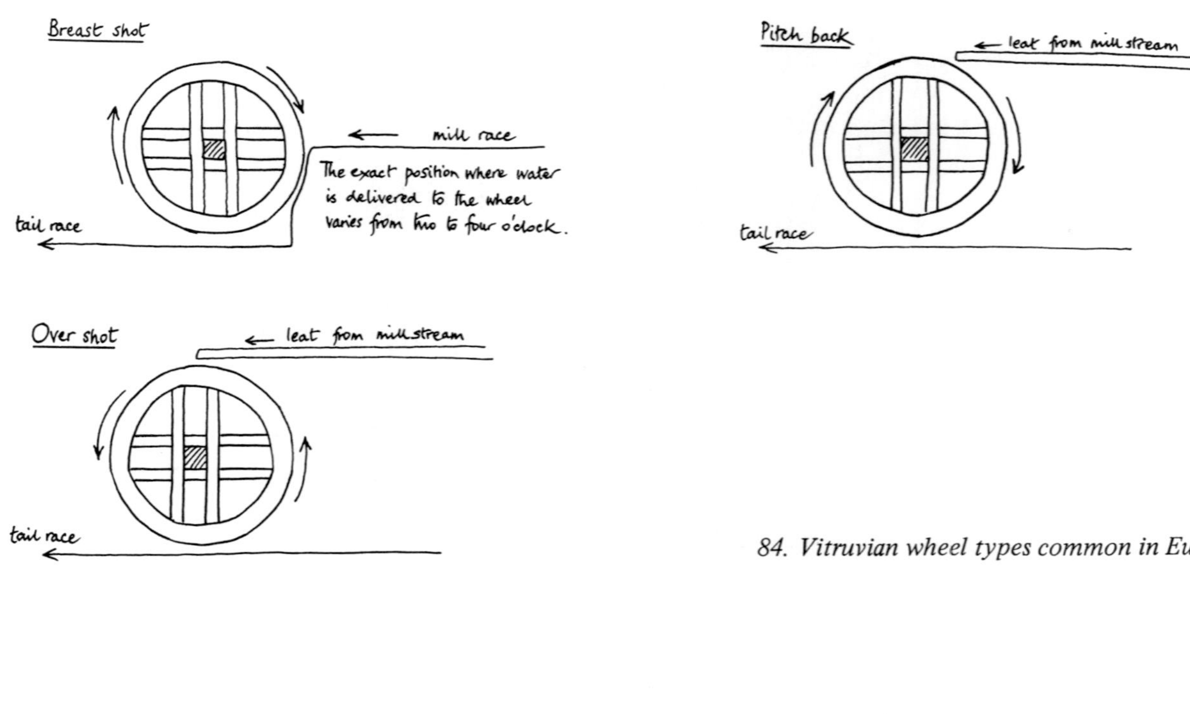

84. Vitruvian wheel types common in Europe.

on the rivers in the Caucasus (Kozmin 1917, 20f). Working examples were seen in 1977 on the south side of the Salang Pass in Afghanistan, where the rather steeper shutes measured c. 4 m (13 ft). However these mills cannot depend upon a constant water-supply and grind slowly. The solution employed in Iran is to tap a perennial source of water and to increase efficiency by utilising the pressure obtained within a penstock rather than merely the kinetic potential of this water. A similar improvement in the performance of Vitruvian mills is obtained in various parts of the world by 'bucketing' the water: it fills long, shallow troughs on the perimeter of the wheel rather than merely striking flat blades in its free fall, as in Iranian examples (see P. 85).

Horizontal mills driven by rivers are still working beside the dams at Shushtar and Band-i Amir (see P. 83). Some mills tap streams directly, as is the case with the Doshman Ziari example in figs 82 and 83, but this is only possible in the foothills of a mountain range, since evaporation naturally dooms streams to a short life on the Plateau. Most mills on the fringe of the central deserts are fed by qanats. Settlements have grown up in connection with the agriculture made possible by a qanat and a mill can be readily supplied from a leat leading off from the qanat outlet. Depending on the lie of the land the tail-race may require a lengthy qanat of its own, as is the case at the deeply sited mill at Salameh, near Khaf. The construction of a watermill, which uses and then passes on for irrigation all water channelled to it, therefore requires skill and may need the capital outlay and disposal of water-rights that traditionally only a landowner could afford.

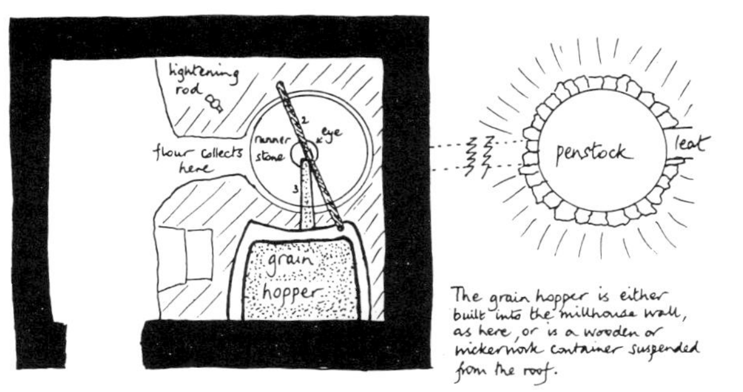

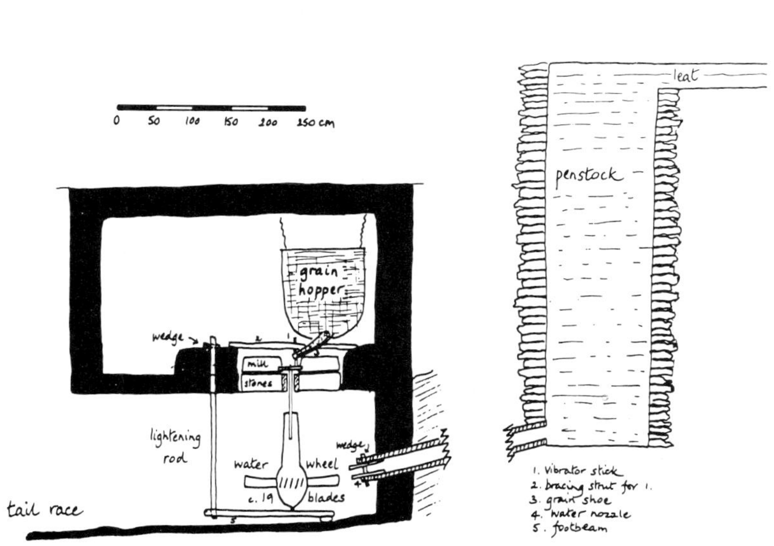

85. A typical horizontal mill in Iran. (top) plan. (bottom) section.

86. Penstock of the mill at Murghab, near Pasargadae 1963 (above).
87. The same penstock viewed from above (left).

Mills are also known within the qanat network, i.e. with both head and tail races underground for considerable distances: one was recorded at Gusheh, near Damghan, in 1968. A straight drop within the general qanat gradient not only adds to the engineering problems of the *muqanni* (see Chapter III), but threatens to erode the base of the channel and entails frequent maintenance in a very confined subterranean space, unless the areas of turbulence at and below a mill and the drop itself are firmly contained within masonry tunnels or baked clay hoops (*kaval*). Few qanats provide the 10 cu.ft (283L)/sec flow which has been reckoned to be enough to turn a mill (Noel 1944, 192), but much more important than the flow of water within the normal qanat gradient is of course its fall at the mill.

A penstock enclosing the head of water

The most developed horizontal mills incorporate a penstock (fig 87), more like a well than a chute, to build up an enclosed head of water, which imparts a far stronger jet of water against the blades of the millwheel than is possible in the system relying merely on the free fall of the water. Known as the *arubah* penstock in Palestine, where it possibly originated, its range and efficiency there have been carefully documented (Avitsur 1960, viii) and the figures confirm that an average qanat flow, when combined with penstocks of a height measured near Kashan (c. 6m /20 ft), Salameh (c. 10 m /32 ft) and Doshman Ziari (c. 4m/ 13 ft) suffices. One notable superiority of the Persian penstocks is their supply from generally perennial qanats, whereas most Palestinian ones fed from mountain streams, stood dry and useless during the summer months.

Such a system has important advantages over a Vitruvian one: small discharge of water needed, no friction loss via gearing, an efficiency of at least 40%, simple construction and repair, and the wheel's force independent of its dimensions (i.e. breadth of paddles or size of buckets on a Vitruvian wheel). They do not however match the hyperbole of the miller's sons at Fin near Kashan who insisted that his (horizontal) mill would grind 2,000 kg (2 tons) per hour! No wonder that their father was "away at home resting" and that the runner stone was raised for dressing! Penstocks in Iran are c. 1.5 m (5 ft) wide and usually either sunk into the ground like wells or jutting out like a stone tower from the hillside just above the mill (fig 86). They are usually mortared on the inside, though the two at Salameh were made of baked brick and incorporated footholds to facilitate any repair work. The present Anglican Bishop in Iran recalls swimming in such penstocks at Taft as a boy fifty years ago.

Crucial to the effective utilisation of the penstock is the nozzle fed by a pipe leading from near its base, which directs water under pressure against the blades of the millwheel underneath the millhouse (fig 85). Depending on the water-supply and affecting the power provided and therefore the rate at which grain can be ground, nozzles of varying diameters are wedged into the outlet of the pipe (fig 88; also Wulff 1966, 280 f.).

88. Horizontal millwheel at Band-i Amir (photo Olive Kitson 1963) (left).
89. View of the grinding area in a mill at Band-i Amir (photo Olive Kitson, 1963). (above).

The horizontal waterwheel

The blades, numbering 18 to 20 in the examples examined, are set into the hub of the mulberry or willow shaft at less of an angle than is common in Europe ; they resemble rough spatulas, whose wider, outer end may be hollowed and all of which may be linked by a wooden or metal rim as in fig 88. True to form and for reasons often reiterated in this book there is little attempt at sophistication in design in order to increase efficiency: these rough-hewn plane-wood blades do the task required of them and are easily replaced by the miller. His access to the waterwheel and the nozzle is up the tail race, generally entered through a trapdoor in the millhouse floor. The penstock is kept empty whilst he completes his tasks, or, alternatively, the pipe to the nozzle may be closed by a shutter.

Simple mill machinery

Once the mill is working the water in the tail race ensures that the steel or stone footbearing which supports the main shaft is well lubricated. As in many horizontal mills in northern Spain a pebble may be placed in the bottom of the main shaft, which bears on a slightly hollowed larger stone fixed into the footbeam. A marginal alteration in the position of the footbeam, achieved by rearranging the wedges at the top of the lightening rod on the mill floor (fig 85), will affect the clearance between the millstones, since the runner stone is supported by the top end of the main shaft. The bedstone is of course set

90. Feed to the stones in the same mill (photo Olive Kitson, 1963).

91. Wickerwork corn hopper in an Afghan mill.

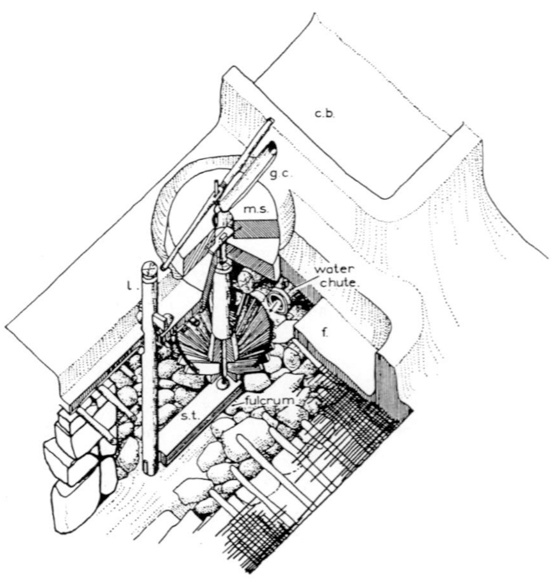

m.s. millstones; c.b. corn bin; g.c. grain channel;
l. lightening rod; f. floor level; s.t. sole tree (foot beam).

92. Axonometric drawing of a mill at Band-i Amir, made by Martin Weaver.

93. Subterranean millhouse at Firuzeh, near Abadeh, 1975.

flush with the floor, while the runner stone is usually surrounded on most of its circumference by a low wall to concentrate the collection of flour at one point (fig 89). A stick, equivalent to the damsel in a European mill, is braced between the edge of the eye and a strut fixed a foot or so above the stones and is attached to the lower end of the feeder channel, which it joggles in order to maintain a constant trickle of corn to the stones (fig 90). Afghan mills often have wooden or wickerwork hoppers slung above the stones (fig 91), but those seen in Iran incorporate the hopper in the general mudbrick structure (fig 89).

The millhouse

The building is almost completely uncluttered, with a total absence of machinery apart from the stones; it is usually dark, being set at least partially below ground level, and for the same reason unprepossessing and easily overlooked from the outside (fig 93). It is difficult to convey the sculptural quality of the inside of the grinding room, where the fine shapes of the plastered mud bins for corn and flour merge with the housing that curves round the runner-stone. This is emphasized by the coating of fine white flour which covers everything in a working mill (including the miller: fig 94), except the spinning focal point of the grey stone and the yellow corn trickling into it. The walls may be dry stone or mud brick, with flat or vaulted roofs (fig 95). Exceptional were the picturesque, tall-domed mills which once flanked pigeon-towers near Herat (Zestovsky 1949, 24).

94. Miller's tea break at Band-i Amir.

95. The grinding area in the mill at Murghab. Of special interest here is the mud brick vaulting: a dome is raised above a square by means of squinch at each corner. 1963.

96. A new millstone for an Afghan mill. It is still wedged on the poplar trunk to which the ox in the foreground was harnessed when the stone was trundled here from the quarry. A team of men had urged on the ox and controlled and braked the stone, weighing about one ton, by means of the longer end of the axle. 1977.

Millstones

Millstones in Persian watermills vary considerably in size, depending mainly on the average water supply. The mill near Kashan, still in regular use, which receives an abundant water supply from the famous Bagh-i Fin, has unusually large stones (155cms/5ft in diameter), quarried like all those used in the Kashan and Isfahan area from the nearby Kuh-i Kargez. Assuming full-time operation such a runner stone should last about eight years, being dressed whenever the millstones cease to crack open and reduce the wheat, barley or millet fed into them (c.f. P. 100 for the dressing of windmill stones). Most millstones are considerably smaller: those measured varied from 93 to 124 cm (3 - 4 ft). They are always in one piece, unlike the much-prized French burr stones used for wheat-grinding in England.

In medieval times millstone quarrying was an important industry at Khullar, N.W. of Shiraz, "a large village where they quarry the millstones used throughout the greater part of the province of Fars, for the stone here is of excellent quality. The curious part is that in all Fars they grind their corn with millstones from this village, but when the people thereof have to grind their own corn they go to some other village to do so, for in their own place there is no stream" (Lestrange 1912, 334f). From Khullar the stones would be trundled slowly and laboriously to their destination on treetrunk axles to which oxen were harnessed, a process still used in Afghanistan (fig 96).

Working hours and output

No reliable figure can be given for the output of Persian water-mills: they work when they are needed, with the minimum of fuss for the miller and to satisfy individual 'small' customers – typical of horizontal milling all over Europe and Asia. One hundred kilos (220 lbs) an hour would be an approximate average, but water conditions may alter that figure drastically. The mill may stand dry and idle for several hours whilst its water-supply is diverted into another irrigation system which must receive its due share at regular intervals. The Doshman Ziari miller (fig 97) whose services were used by between five and eight hundred people prior to the installation of a diesel powered mill at Deh No in 1964 ground a daily average of c. 450 kg (1000 lbs) through nine months of the year. A very rough ratio of one watermill to one thousand people holds good for the cities of Isfahan and Kerman a century ago (Houtum-Schindler 1881, 331 and 1898, 123), enabling one to arrive at tentative figures for the output of mills, and therefore their value, in their heyday: a daily average of perhaps 400 kg (880 lbs), which is very acceptable for typical Persian watermills. By 1976 the Doshman Ziari miller's custom had declined to no more than 20% of the earlier figure.

Mills were formerly a service industry in most villages since bread is the staple in the diet on the Plateau and large quantities of flour are needed. Nowadays with good communications and cheap electricity the requirements of a wide area are readily supplied by mechanically powered mills in the towns. Until the 1950's a donkey-load of wheat would be taken to the mill by most peasant families about every three weeks (one's future bread supply would be stored as grain since stone-ground wholemeal flour does not keep for very long in the hot Plateau climate). The average adult male Persian eats one kilo of bread per day. This represents roughly half an hour's work with a rotary hand-quern (Curwen 1937, 30), a utensil seldom found in any Plateau homes today, though querns of several designs were once widespread (fig 98).

Grinding flour by hand added considerably to the household chores. In an Afghan village in 1964, despite easy access to water-mills: "A young woman – just home from a day's work in the fields – immediately starts to grind flour for baking bread in preparation for the evening meal. The rotary quern is placed inside the door to the right, and in the evening the rasping sound from all these querns fills the village air" (Edelberg 1979, 59).

By contrast, thirty years earlier, a troglodytic dweller in the mountains near Kerman, who one might have expected to use a quern, made a long trek three times a week: "On baking-day Fatima was up at two, to walk several miles across the desert carrying wheat, barley or a mixture to be ground into flour, wholesome but coarse and not very clean. As soon as it was ground, in the dark, subterranean mill run by water, she put it on her back, minus what she gave to the miller and plus the bits of millstone, and was back home soon after seven ... " (Merritt-Hawkes 1935, 183). It would seem that both women might have made life easier for themselves by having several weeks' supply ground in the mill at one time, providing there was a donkey handy for the few hours required. From the point of view of the miller Fatima's small parcels of grain were surely inconvenient? They would not have filled his hopper, therefore demanding his close attention since working stones require a constant feed, otherwise they may be damaged. Perhaps ladies of similar habits pooled their grain at the mill and the ensuing squabbles over the precise amounts of flour due to each were part of their thrice weekly ritual.

Tolls and taxes

The usual miller's toll was one part in twenty of corn ground. The rent he paid to the mill-owner depended upon the likely volume of trade when the contract was drawn up: the Doshman Ziari miller, for example, paid an annual rent of 700 kg (1540 lbs) of flour in 1976. The traditionally depressed and indebted state of most peasants on the Plateau and the dependence of mills there upon qanats meant that a landowner or, less often, a group of small farmers owned the local mill. In western

97. The miller of the Doshman Ziari mill. His grinding area closely resembles that at Band-i Amir (fig 89), though the mechanism controlling the grain shoe is cruder and the flour sweeps out unhindered on half the perimeter of the stones. (1976, Susan Wright).

98. A saddle quern, still used occasionally for grinding lentils. The Doshman Ziari tribeswomen work the heavy stone rubber backwards and forwards. This is a simple form of the ancient saddle quern, used long before the invention of the rotary quern. (1976, Susan Wright)

Afghanistan the construction of a dam for a watermill required the permission of the owner of the irrigation canal and was expensive to obtain, so, not surprisingly, many mills there belonged to qanat owners (Humlum, 1959, 220, 235). In early Muslim times mills at Qum paid less than half the normal tax to the state if they were turned by water from a qanat rather than a river, because of the cost of maintaining the qanat (Houtum-Schindler 1898, 71). At the beginning of this century in Khorasan, at a time when the poll-tax stood at 2 tomans per family, "every watermill pays from one toman to twenty tomans a year or even more as taxes in proportion to the income of the mill owner" (Sykes 1910, 5). This was simplified in 1926 when land taxation became uniform and "mills were to pay 5 per cent of the owner's receipts" (Lambton 1953, 183).

Unusual rights existed in certain mills. During the twelve days after New Year's Day the local *Seyyeds* (descendants of the Prophet) still receive the proceeds (i.e. the rent for that period) from the Doshman Ziari mill. At one Afghan mill in a village west of Kandahar (Pirzadeh) the miller received flour ground every Thursday night as his due from the peasants whose share in the mill entitled them to free milling otherwise (Humlum 1959, 222).

A final comment on past practice, which implies that water-milling was once reckoned to be a profitable business: "Mr. Bruce informed me that there still exists in Persia an ancient custom in cases of emergency, of requiring from every mill (the wheel of which is turned by water) a man and a horse armed and accoutred for the field; and of these they reckon one hundred thousand" (Morier 1812, 240). The Persian army may thus have found itself swollen at times by 100,000 unpredictable cavalrymen, more like Don Quixote on Rosinante than Rustam on Rakhsh, a situation described by the same James Morier in "Hajji Baba of Isfahan" (Morier, 1897, 206 f.).

Mills fed by dams

99. Watermills at Shushtar. The dam, Band-i Gerger, behind the photographer, provides a powerful head of water for numerous mills on both sides of the canal, as can be judged by the torrents gushing out below these mills. (1971).

The best-known mills operating beside a large dam which provides a good head of water are those at Shushtar in Khuzestan (fig 99), where the dam is tapped part way down its great height to feed the penstocks of the mills (Curzon 1892, 373). Before the Second World War there were thirty mills here, each grinding on average rather over 1000 kg (c. 2500 lb) per day, thanks to this 9.5 m (31 ft) head (Lockhart 1939, 89).

On the Plateau itself there are the watermills clustering at one end of the Band-i Amir in Fars (fig 100). This dam was one of the wonders of south Persia in medieval times. From the description by Muqaddasi, a contemporary of the dam's tenth-century builder, it would seem that the mills were not originally operated by penstocks as today, but in conjunction with norias (see Chapter III): "By a mighty dam the water was raised into a lake. On both sides of the lake he (Adad-ud-daulah) set up draw-wheels driven by the water; there were ten of these; under each was a mill. He thus irrigated 800 villages by the aid of pipes" (Mez 1937, 451).

A caustic comment by an English traveller who visited Band-i Amir in 1851 not only implies mills on the present pattern but also finds little else to praise: "More than 20 small watermills are situated under, and at the sides of, the *bund* (dam), keeping up an incessant humming noise, the only sign of industry in the place. The people here seem to be a stupid, witless set. With the fine waterpower they possess – and in this respect they are better provided than in any other place I have seen in this country – they might have gardens, vineyards ... but for their indolence and apathy" (Binning 1857, 366).

The penstocks here, providing a 4 m (13 ft) head of water, can be seen from figs 100 and 101 to be part of the millhouses, attached like amoebic parasites to the side of the dam. The dozen mills still operating in 1963, using a steady water supply and in proximity to an extensive area of cereal cultivation, represent an exception in the normal pattern of horizontal watermilling in Iran: an enterprise that could be run on a large commercial scale. Depending on the decree of the original benefactor the mill rents in such cases might contribute to the upkeep of the dam or be administered as *vaqf* (going to charitable or pious purposes), rather than line the pockets of his heirs.

It is an abundant water supply which encourages a highly profit-conscious approach to milling. This was seldom possible on the Plateau, but characterised the mills concentrated on the rivers just west of the Zagros mountains: at Dizful, Shushtar and, most magnificently, Baghdad. There the Caliph al-Mahdi provided the financial backing for a Byzantine ambassador to build and run what became known as the Mills of the Patrikios. These originally contained a hundred millstones and produced an annual income which made the ambassador a millionaire and led the Caliph to annex them on his death c. 780 A.D. (Lestrange 1900, 42). This great state-controlled enterprise had vanished by 1300. Since it was powered

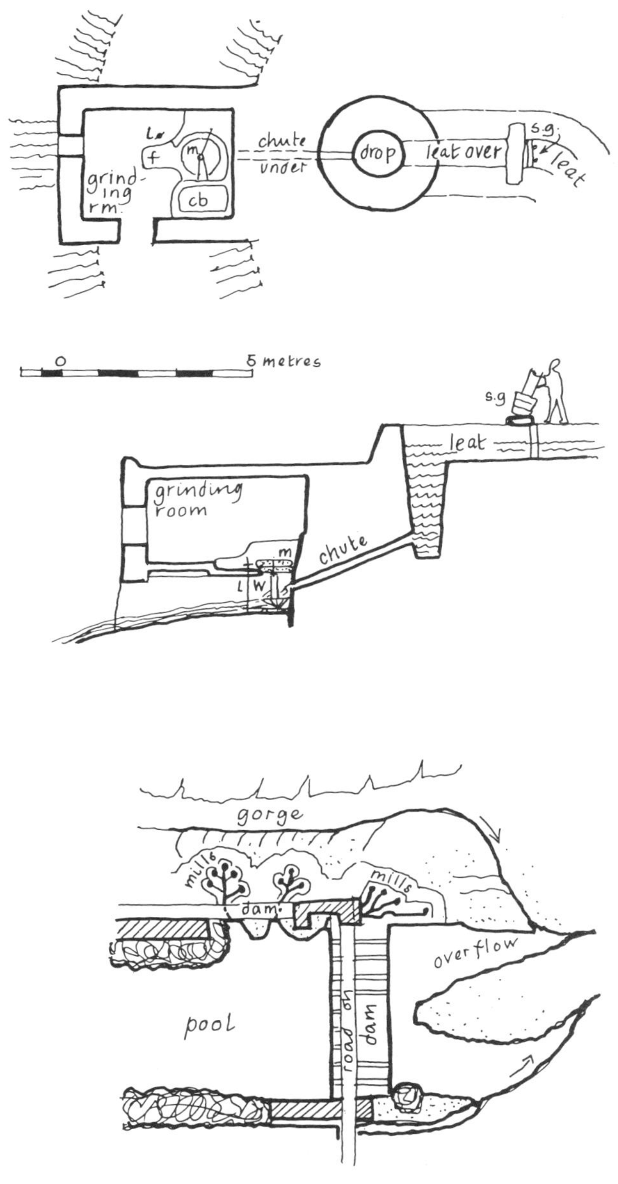

100. Diagrams illustrating the mills at Band-i Amir.

101. Leat and penstock heads at Band-i Amir (Olive Kitson 1963).

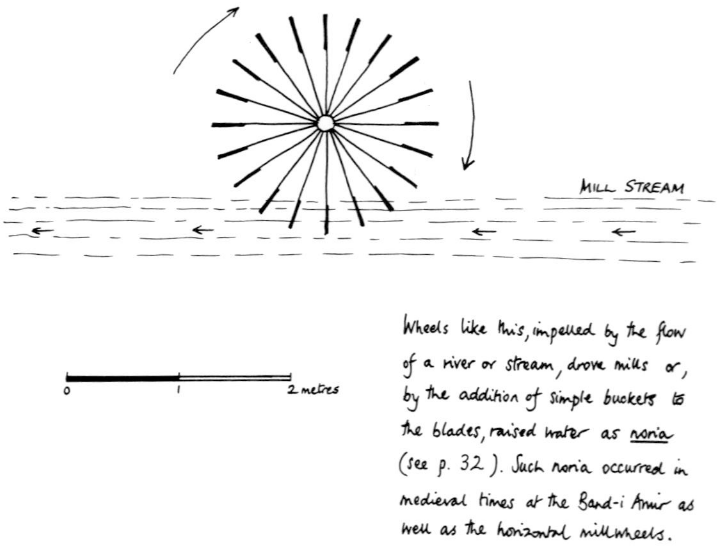

102. An undershot wheel.

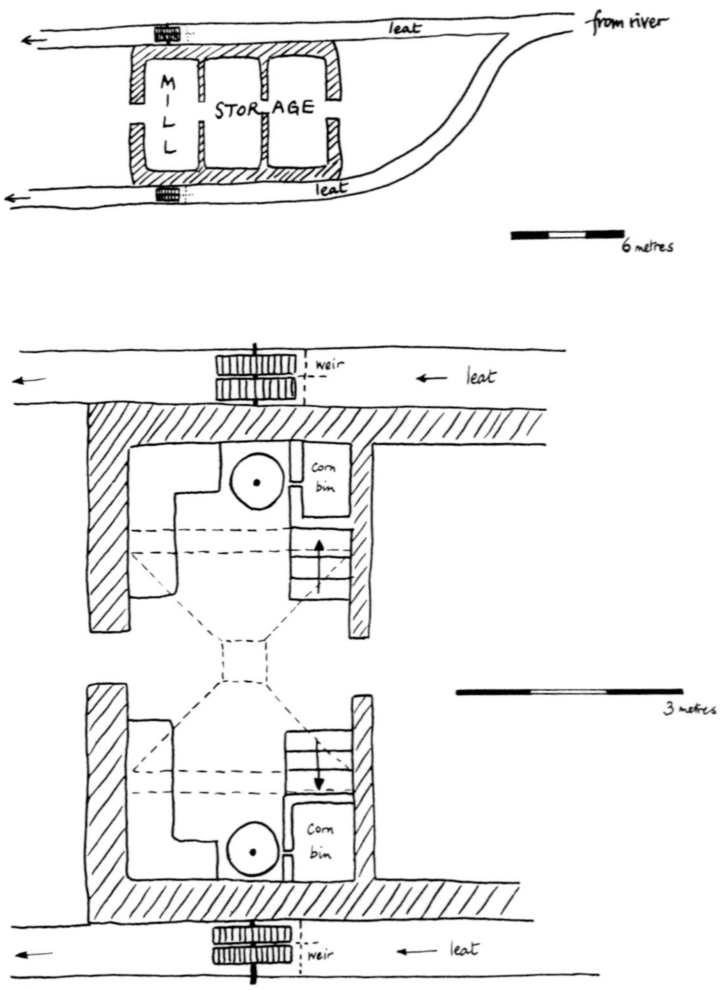

103. The mill at Pir Bakran: The position of the millhouse (top). A plan of the grinding area (bottom).

by the river Tigris one might assume the wheels to have been Vitruvian rather than horizontal. There is no indication either way and both historically and technologically either would be appropriate.

Ship-mills

The Vitruvian principle of a vertical mill-wheel driving the stones via gearing attached to a horizontal main-shaft was utilised in Mesopotamia from early Muslim times. Mills erected on ships moored in the Tigris are described at Basra by Muqaddasi and at Mosul by ibn Hauqal, and on the Karun at Ahwaz by Yaqut. At Mosul "each mill had two stones, each of which ground fifty camels' loads per day" (Mez 1937, 466), a prodigious output (c. 18,000 kg/ c. 40,000 lb per mill) compared t that of the horizontal mills supplying the needs of village communities which we have been discussing hitherto.

These floating mills, with the waterwheel supported between two boats, one of which contains the grinding me anism, require a large river flowing at a moderate speed; they are liable to flood damage and the boats need frequent maintenance. They were still in use on rivers in Romania until recently. No rivers on the Iranian Plateau can ever have had sufficient depth and perennial flow to justify the introduction of such mills from the Mesopotamian part of the Abbasid caliphate.

Vitruvian mills on the Plateau

The only river on the Plateau which provided sufficient water to power Vitruvian wheels in conventional millhouses throughout the year was the Zayandeh Rud. Despite the greatly increased demand on its water for industry and agriculture in the Lenjan valley above Isfahan and the recent damming of the river just below Isfahan, a few still operate. "Watermills are numerous on the banks of the aqueducts (*jub*, see Chapter III) and their wheels are almost invariably undershot" (Binning 1857, 117; fig 102 illustrates a simple undershot wheel). Over a century later one has to search har to locate a watermill working off the Zayandeh Rud. There is one in Isfahan near the Pol-i Marnan and anoth some way upstream at Pir Bakran, but those near the ol Pol-i Shahrestan have been submerged by the lake behin a modern dam. These mills have a breast-shot wheel on each side, so that the mudbrick, domed millhouses stand on artificial islands. Each wheel drives its own pair of stones and when business is brisk and the river is reasonably full, as in spring, both wheels operate simultaneously (fig 103).

The mill at Pol-i Marnan was grinding barley for ca tle-feed in May 1977, but the low level of the river mea that the miller could only use the smaller of the two waterwheels. His output was therefore considerably lower than the 180 kg. (c. 400 lb.) per day he reckoned to grind when the larger wheel could be brought into operation. This latter was simply two wheels (of diameter c.2m /6ft) placed side by side on the shaft, a well-known device for increasing power when the mill is breast-shot and when the wheels are constructed solely of wood (fig 106). There is no attempt to bucket the water in these Persian mills: the blades are straight pieces of willow or plane, no more than 50 cm. (1 ft 8 ins) in length, fixed to the twenty pairs of spokes which are themselves linked at the rim by spacing struts held in place by wedges (fig 105). Such ramshackle affairs cannot be allowed to revolve at a great rate or to cope with fast-flowing water: they would simply disintegrate. The miller makes and mends his own machinery. The gearing is all wooden: the 36 apricot wood cogs in the pitwheel mesh with the lantern pinion, which needs reconstructing about every fortnight of constant use, using hardwood from the Caspian forests (fig 104) The stones, measuring c.1m (3ft 3ins) in diameter revolve six times for each revolution of the mill-wheel.

The wooden working parts that require frequent maintenance, the low capacity of these mills and their vulnerability to the reduced flow in the Zayandeh Rud over the last decade, as well as the ease and speed of transport between the villages and Isfahan's electrically driven flour mills, all conspire to bring about their extinction sooner than the simple horizontal mills in remote areas like that described at Doshman Ziari in Fars.

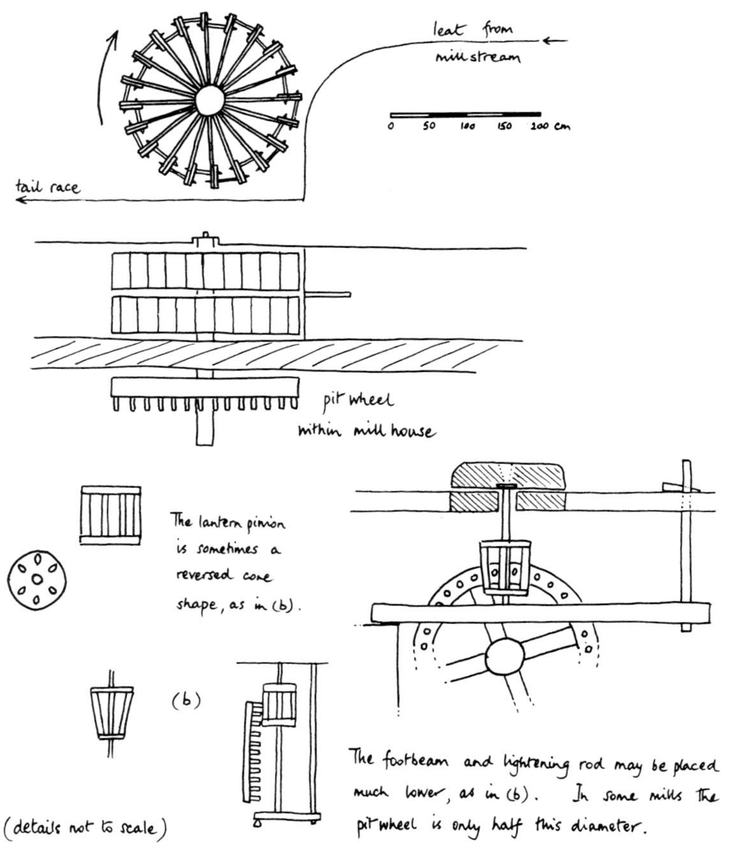

104. A Vitruvian millwheel at Isfahan, with details of the gearing.

106. Vitruvian wheel at work.

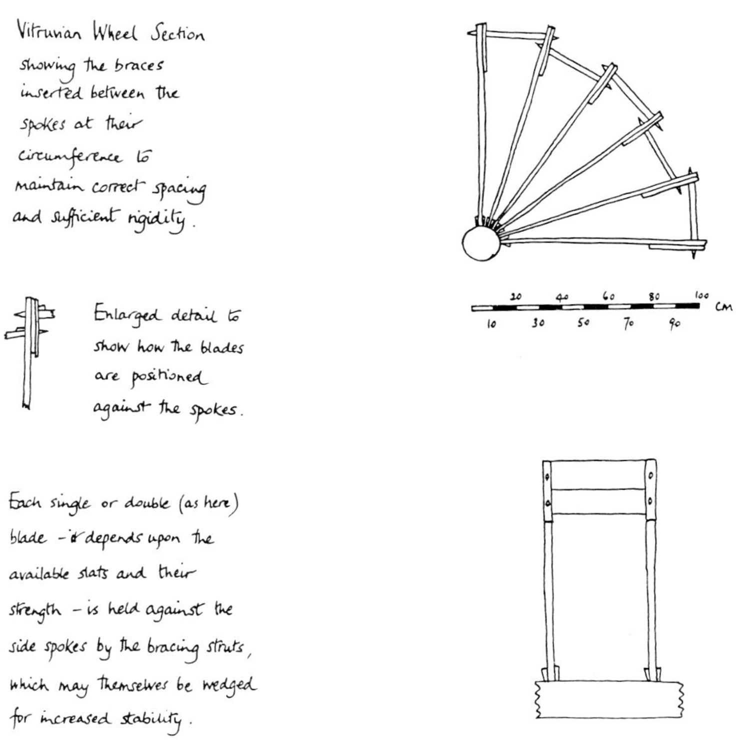

105. Details of a Vitruvian wheel.

107. Vitruvian millwheel at Pol-i Shahrestan, outside Isfahan. (1962).

A fantastic medieval mill

One of the strangest watermills recorded in Iran was at Marand, N.W. of Tabriz, in al-Dimashqi's day (c. 1300). It was Vitruvian and particularly drew his attention because of its independence of a constantly renewed source of water and because it seemed to have solved the question of perpetual motion. His detailed description still leaves one asking important technological questions and he does not say how efficiently it ground corn. But for al-Dimashqi this mill was one of the wonders of the world, the age and civilisation. The millhouse contained a deep reservoir, above which were suspended two waterwheels, each connected to a pair of stones; water piped up from the reservoir was directed onto the blades of one wheel and was then sucked up and funnelled across to turn the wheel on the other side (Wiedemann 1970, 214 f.). The pumping mechanism, about which al-Dimashqi remains silent, could have been animal-powered like the water-raising device referred to in Chapter III or the grinding mills that still employ camels in the Isfahan bazaar and at Herat. The enclosed water system is an intriguing answer to the evaporation experienced on the Plateau, but the mechanism sounds too complicated for any except a wealthy landowner to have employed and maintained.

It sounds in fact rather like the water-powered automata devised by al-Jazari to entertain the princes of Diyarbekr (in modern Turkey) c. 1200 A.D., which indicate a good understanding of valves, gears and the crank. The waterwheels in his devices are mainly horizontal, known by him and earlier Arab mechanical engineers as *rumi* (Byzantine, i.e. Greek) despite their equally old and probably independent existence on the Iranian Plateau. His gearing in certain examples however is reminiscent of that operating in the watermill at Isfahan described above: individual cogs protruding like right-angled pegs from the pit-wheel and turning a lantern pinion (Hill 1974, 275).

The neglect of inventive technology

The technology behind al-Jazari's clocks and wine-dispensers disguised as peacocks is sophisticated for his time and the wonder is that none of the admirers who copied his book were moved to attempt to apply these ideas to useful machinery. As Lynn White writes in his foreword to Hill's edition of al-Jazari: "Why, for example having taken over paper-making from the Chinese, did

Islam – which used waterpower for grinding grain – never (so it seems) adopt the Western use of water-power for pulping the rags from which paper was made? Why having invented the world's first operative windmills in Sijistan, did Islam fail to spread them more widely?"

Rice-mills

The waterwheel that ground corn in Iran never became a prime mover for other agricultural and industrial processes there, as it did, so importantly, in medieval Europe. It was never developed by engineers, so inventions were not made to ease the burden placed on human and animal muscle, and the pre-industrial age lasted well into the twentieth century. An exception are rice-hulling mills (*abdang*) formerly driven by under-shot wheels in the low-lying Caspian provinces and by mountain streams diverted down hollowed-out tree trunks in the Alborz valleys. The water strikes narrow blades typical of a horizontal millwheel, but set vertically on a horizontal shaft which works the rice-hulling pestle by means of a cam (Ferdinand 1959, 210). Such an application of water-power may have been introduced in Mongol times from China. They occur across a wide geographical arc of Asia, but within our area principally in Afghanistan where rice is the staple of life. On the Plateau, where bread is the real staple and rice more of a luxury, rice-mills have been recorded in several towns (Ferdinand 1959, 219), including interestingly Jiruft, S.W. of Bam, in 1959, where a thousand years ago Ibn Hauqal found 50 watermills driven by the Devil's River (Mez 1937, 466).

Small, slow and safe

There was indeed little chance or incentive to develop the uses of water power on the Plateau in medieval times. The invasions of the Mongols and of Tamerlane not only destroyed irrigation systems nurtured for centuries and decimated the population, but also discouraged the survivors and succeeding generations from thinking of more than getting in the next harvest safely. These naturally parochially-minded, isolated communities, scattered over the inhospitable Plateau, were content if their simple mills ground enough flour for their needs. They had no wish to stimulate any flourishing business which would profit only the landowner to whom they were already heavily financially indebted, or bring the Shah's tax-collectors looking to increase the assessment. The terrifying impact of earthquakes, whose incidence on the Plateau has been again most marked during this last decade, with their utter devastation of settlements, qanats and water-systems, was yet another nail driven into the coffin of Persian inventiveness. The peasants were resourceful and resilient, but they had no taste for venturing their meagre capital in new machinery, for giving it a chance to prove itself and for making adaptations to suit their own conditions (cf. Chapter VII, P. 96). The horizontal watermill satisfied their need for a reliable means of grinding small quantities of flour in a country where the water supply is limited. It was, in this context, efficient. Only with the present availability of cheap electricity and diesel fuel has it fallen into disuse.

It is sad that the Persian's awareness of waterpower did not encourage him gradually to develop and come to terms with at least some of the many industrial processes which the new sources of energy have now brought to his country at bewildering speed.

108. A pair of Afghan mills near the Iranian frontier, viewed from the south-west. See also fig 113. Notice the stepped and curving side walls and the pierced decorative brickwork of the east wall. (photo Richard Hewer, 1971).

7 WINDMILLS

To the traveller coming from the north the small town of Neh in eastern Iran appears to be defended by a high, unbattlemented wall with surprisingly deep and wide arrowslits at regular intervals. From the other side it is clear that this long wall has over sixty separate compartments at first floor level, each containing a tall wooden spindle that meets a solid crossbeam at the top. These are the substantial remains of a row of horizontal Persian windmills which once supplied most of the flour for the several thousand inhabitants and for the caravans of merchants and pilgrims moving to and from the city of Mashhad along this main eastern artery of Iran. Neh is roughly equidistant from Zahedan and Birjand, about ten camel days in each case. The country is broad and barren with very few settlements to the south and east where the road seemingly divides the Dasht-i Lut from the Hamun-i Helmand on the map. From the north, where the mountains soon begin to rise steeply, comes the summer wind which powers these mills, unique in their design to this part of the Plateau.

Five or six tiers of wooden spokes slotted into the central spindle support vertically aligned slatted or reed blades at their extremities. These catch the wind as it blows through the slit in the front wall and twirl like the revolving doors of a hotel entrance, causing the spindle to turn the runner stone immediately below in the millhouse at the same speed. Such a mechanism closely resembles a waterwheel turned into the air so that it may be powered, not by a steady jet of water, but by the inconstant gusting of the wind (fig 110).

These windmills are only found in the path of the 120-day wind, which blows from the Kara Kum in Turkmenistan to the Baluchistan desert during the late summer months. It affects a 'corridor' extending across the lines of longitude 60° – 62°, not much more than a hundred miles, including the border between Afghanistan and Iran. The windmills are situated within a three hundred mile long stretch of this 'corridor', from the Bakharz mountains and the Hari Rud valley in the north to the Hamun-i Helmand in Sistan in the south. As Ibn Hauqal commented when he visited Sistan a thousand years ago: "The ceaseless wind was used by the people to turn their windmills which were a feature peculiar to this country" (Le Strange 1905, 337).

There are no reliable records of such windmills beyond these boundaries. Certain puzzling, scantily documented references to windmills in Yazd, Kerman and

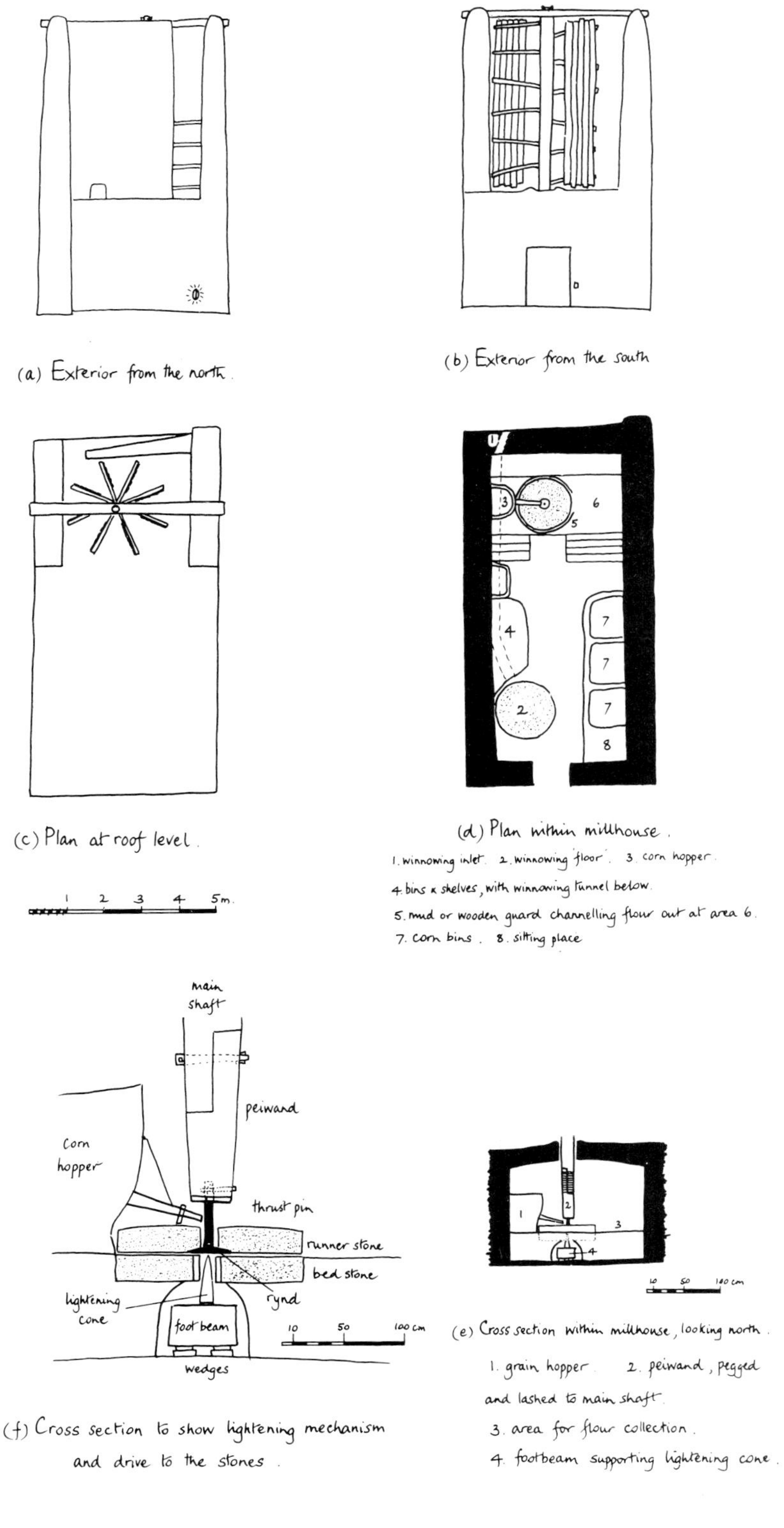

109. Diagrams to illustrate the layout of a typical Persian windmill.

110. The wind-wheel of a working mill at Nishtafun, 1977.

Bam exist, but they seem likely to have been isolated experiments which soon failed – for the same reasons that these windmills were not built elsewhere on the Iranian plateau: insufficient or unpredictable wind and a tradition of satisfactory watermilling. The 120-day wind blows only from the north, a fact which has determined the design of these structures, as will be seen. Elsewhere on the Plateau the wind is much more variable in intensity and direction, so these windmills would be both inefficient and inappropriate (for one such failure at Kerman a century ago see Floyer 1882, 321f.).

Structures

Rows of horizontal windmills are a striking feature of several villages near Khaf in Khorasan and further south at Tabas Masineh and across the Afghan border at Anardarrah (Grötzbach 1974, 114f). They stand on the northern outskirts of each village, presenting a common front to the wind, as at Neh. Each mill's front wall is in fact often slightly angled towards the slit, encouraging the wind to find the gap. Behind the front wall and rising to the same height is the wind-wheel, whose top bearing, a ring of apricot wood, is set into a heavy beam of mulberry or a section of superannuated windshaft resting directly on the side walls. The wind-wheels are clearly visible from the south, for there is no rear wall on these mills in Iran.

111. Derelict mills at Neh, viewed from the south in 1977. An almost identical photo was taken by Sven Hedin in 1904 during the mill's active life, but in early April before the spokes and reed mat sails had been refitted for the grinding season (Hedin 1910, fig 237).

The millhouse is built of sun-dried mud brick, with the usual mud plastering. The walls surrounding the wind-wheel taper from 90 cm (c. 3 ft) at roof level to 50 cm (1ft 8 ins) at the top. The north-east corner of each partition wall is occasionally strengthened by thickened courses at this point or even by a branch laid across the top angle. Sometimes there is a passage through the mills at roof level if the row is a long one — four hundred metres for the long terrace of mills at Neh — so that an individual miller can get to the front of his mill quickly if necessary. This will be to screen off the wind-slit, to which access is gained either by steps up from the ground or by crawling through a low hole near the corner of the front wall. North doors are unusual, probably because any draught would add dust and confusion to the working area beside the stones (fig 109).

There is however a small aperture, barred to discourage thieves, at the bottom north-west corner of most windmills in the Khaf area. This leads the wind through a long tunnel which emerges within the mill, not far from the south-facing door. The miller can then winnow the grain in comfort and in the shade. Or he can rest in the cool flow of air: 'this is my air-conditioning system' remarked one miller.

The mill may be roofed by the mud brick barrel vault common in houses in Eastern Iran, or a series of two or three domes (fig 112). The opening for the wind-shaft in the bay over the stones has a raised rim to keep out the dust to some extent and to provide a firm support for the retaining peg when the whole wind-wheel is raised to free the runner stone for dressing (see P. 99). At this point the roof is about 50 cm (1 ft 8 ins) thick.

Rows of mills

Mills are built in terraced rows to economize on space and building materials, to improve stability and security and for the same clannish reasons which cause all the leatherworkers, the coppersmiths or the weavers to congregate in their own sections of an eastern bazaar. One miller, however, stated from experience that a free-standing windmill in his village had generated 50% more power from the wind than the one he was now operating in a row of seventeen. Its detached isolation had not lasted long. Tolerable efficiency is all that most peasants ask for. There is sufficient wind in the late summer months, and as one smug local saying boasts, 'Khaf takes its toll of the wind'.

Exceptions to this herd instinct are common in Afghanistan, where semi-detached and solitary mills are common. This may indicate a more independent spirit or a more specialised miller's role than exists in Iran, or it may result from an awareness that the wind here is blowing less strongly on the edges of its geographical

112. Mill roofscape in Nishtafun, 1977: dome, barrel vault and domes.

113. A pair of Afghan mills viewed from the north. Fig 108 is another view of the same mills. (photo Richard Hewer, 1971). Notice the deep angling of the front walls towards each wind-slit, the blind arcading at ground floor level and the absence of inlets for winnowing passages.

'corridor'. An additional argument for the latter interpretation is the fact that Afghan mills have a stepped and curving eastern wall which encloses much of the rear side of the windmill. The turbine effect thus created must enhance the efficiency of the windmill (fig 108). Yet the millers of Khaf, barely a hundred miles away to the west, knew nothing of this refinement and expressed no interest in adopting it. Once again, their traditional mill design suited their local requirements. Gabriel, who was exposed to the rigours of the climate of Khorasan for long periods in the 1930's regarded Khaf as the windiest part of the whole region (Gabriel 1935, 165).

Local materials

It comes as a surprise in an arid region to discover that all the materials for the wind-wheels are available in the villages. The 7 m (c. 23 ft) long windshaft, the spokes and the slats which make up the sails are pine, the crossbeam is either pine or mulberry, the lightening cone is made of ash, and the footbeam is turpentine wood. Stands of pine trees are in fact common wherever water is brought to the surface in this north-east part of the Iranian Plateau. The sails sometimes consist of bundles of reeds in certain villages, also grown locally; however the weight and bulk of these thickly bunched reeds must increase the inertia and uncertain balance of the wind-wheel (fig 114). Slats of pinewood appear to be more suited to the purpose, however haphazardly they are attached to the spokes. A network of wires, salvaged from old lorry tyres, holds the whole wind-wheel together (fig 115). It should be mentioned here that the reed mats photographed by Ferdinand in his excellent account of Afghan windmills were much more neatly tailored than those seen near Neh (Ferdinand 1963, 84f.).

114. Reed mat sails at Chahar Farsakh, near Neh. Notice the crooked spokes, the jointed wind-shaft (unusual) and the ramshakle rigging, 1971.

Maintenance

The English windmill has always been vulnerable to fire and the weather. Deterioration of wooden structures and machinery if maintenance work is not carried out regularly is disturbingly rapid in our damp climate. By contrast little needs to be done to keep a Persian windmill in working order providing the original structure was soundly built. Earthquakes, flash floods, and winter snow are the occasional wreckers of mud brick buildings. Apart from neglect, the long-term enemy of the windmill is its own prime mover, the wind. Wind-erosion has scoured the ruins of the cities of Sistan until they have lost most of their detail and much of their superstructure; they include the substantial remains of the "chigini" (decorated) windmill probably built in the 17th century (Tate 1910, 160 illus.). Carefully laid mud brick is a very durable building material, so little maintenance work is needed on the windmills beyond adding a fresh coat of mud plaster to the roof every autumn. The owners do not even trouble to lay a protective course of baked brick tiles on the top of the walls. The sails and spokes are sometimes dismantled for the winter, but normal practice is to leave them *in situ.* That way the occasional rain will swell the pegs and joints temporarily, before the next dry spell restores the element of 'give' under which the device operates best. Nothing is tailor-made about the wind-wheel: it is no miracle of the joiner's craft. It is an assemblage of handy branches and off-cuts from the local carpenter, within the capacity of any industrious and patient man who knows what is needed by looking at his neighbour's successfully operating wind-wheel.

115. Close up of a windshaft at Khaf, showing the pegging of the spokes, the retaining wires and (top right) one means of tautening the framework

Origins

Persian windmills are at least five hundred years older than their European counterparts. The two inventions, technologically quite different, were made independently and neither has gained ground at the other's expense. The glib statement in some history books that the Crusaders brought back windmills from Palestine is without foundation; more likely is the assumption that they acquired the notion of grinding corn by means of the wind. Imprisonment in Iran, contact with Persian mercenaries in Muslim armies, with Arab geographical works, even with the Assassins whose order originated in eastern Iran near Qain, provide plausible opportunities for Crusaders hearing of windmills at second hand. Then in late twelfth-century Europe millwrights and engineers produced their own machine to translate this notion into reality.

The earliest reference to Persian windmills is for 644 A.D. in early tenth-century encyclopaedic works by al-Masudi and al-Tabari. A Persian slave Abu Lulua, complained to the Caliph Omar in Medina that his tax assessment was too heavy:

"And what is thy trade?" asked the Caliph. "A carpenter and worker in iron", he said. "It is not much", replied Omar, "for a clever artificer like thee. I am told that thou couldest design for me a mill driven by the wind." "It is true." "Come then", continued the Caliph, "and make me such a mill that shall be driven by the wind." "If spared", said the captive in a surly voice, "I will make a mill for thee, the fame whereof shall reach from east even to far west"; and he went on his way. Omar remarked, as he passed, the sullen demeanour of Abu Lulua. "That slave", he said, "spoke threateningly to me just now" (Muir 1924, 187). Sure enough, the following day Abu Lulua murdered Omar and was himself hacked to pieces by the guards. So the famous mill was never built and the Arabs showed no further interest in the device. It is revealing that Abu Lulua was a carpenter and a smith by trade. The tone of Omar's request suggests that the invention of the windmill was probably Persian (certainly not Arab) and predated the Muslim conquest.

Other Persian and Arab geographers add a few details of uses and sites, naming Sistan and the Hari Rud valley (the former much more fertile than today) as the home of these mills. No claim is made as to the antiquity and origin of the invention, though al-Mustawfi indicated in the fourteenth century that these windmills were popularly ascribed to the Pharoah of Egypt at the time of Moses who had once campaigned as far east as the site of Herat (Le Strange 1905, 411). This legend, in which Ramesses II becomes a much-travelled "Fasael", was still current amongst illiterate millers in one Khorasani village in 1977.

A medieval description

The earliest description of both the architecture and the detailed machinery of Persian windmills occurs in the writings of al-Dimashqi c.1300:

"When building mills that rotate by the wind, they (in Sistan) proceed as follows: They erect a high building, like a minaret, or they take the top of a high mountain or hill or a tower of a castle. They build one building on top of another. The upper structure contains the mill (*raha*) that turns and grinds; the lower one contains a wheel (*doulab*, meaning "scooped water wheel") rotated by the enclosed wind. When the lower wheel turns, the upper mill stone turns too. Whatever wind may blow, the mills rotate, though only one stone moves. After they have completed the two structures, as shown in the drawing, they make four slits or embrasures (*marma*, meaning "loophole of a fortress") like those in walls, only they are reversed, for the wider part opens outward and the narrow slit is inside, a channel for the air in such a way that the wind penetrates the interior with force like in the case of the goldsmith's bellows. The wider end is at the entrance, and the narrower end on the inside so that it is more suitable for the entry of the wind, which penetrates the mill house from whatever direction the wind may blow (hence the four openings in the structure). If the wind has entered this house through the entrance prepared for it, it finds in its way a reel (*saris*) like that on which the weavers wind one thread over another. This machine has twelve ribs (*dil*), one could diminish them to six. On these, fabric (*cham*, meaning "rough unbleached linen") has been nailed, like the covering of a lantern, only in this case the fabric is divided over the different ribs, so that each single one is covered. The fabric has a hump which the air fills and by which they are pushed forward. Then the air fills the next one and pushes it on, then it fills the third. The reel then turns, and its rotation moves the mill stone and grinds the corn. Such mills are wanted on high castles and in regions which have no water but a lively movement of the air " (quoted in Wulff 1955, 285f.).

116. Medieval illustration of a Persian windmill.

117. Derelict single mill outside Herat.

Annotated drawings accompany this account in both the earliest surviving manuscripts (fig 116). In three important respects the windmills he describes differ from those observed by nineteenth and twentieth century travellers: the millstones are above the sails, the sails are made of material (linen) and the building is constructed to take advantage of winds from all points of the compass. The placing of the stones in the natural position for a horizontal watermill never occurs again in references to these windmills. It was perhaps a logical first step, but the advantages of more convenient grain feed, access to the stones and raising the wind-wheel higher must have soon led to the reversal to the now universally accepted position. The shrouding of the wind-wheel on all four sides, with slits to welcome wind from all points of the compass, is another facility otherwise unmentioned, except in Chinese sources about Herat and Samarkand (Needham 1965, 560); the reasons for the terraced rows facing northwards becoming standard have been rehearsed here already. The comparison with a weaver's reel is very apt if one has seen those still used in the bazaars of Herat; whereas the few windmills there stand idle and derelict (fig 117).

European observers

The passing references to windmills in late medieval histories of Sistan indicate that they were an accepted part of the village scene. The lengthy gap in Marco Polo's account of his travels coincides tantalisingly with the very province of Khorasan where we might expect him to have seen and commented upon such weird contraptions, for the windmill was probably unknown in Italy as early as 1271. The first European to mention seeing Persian windmills was von Poser c. 1620, in Birjand (Gabriel 1952, 59).

There are numerous references to the windmills of Khorasan and Sistan in nineteenth century travellers' diaries, though none of them devote much space to their operation or to their place in the rural economy. Most comment on their unusual design, so unlike the tower mills then at the height of their importance in Western Europe. However, one Englishman travelling home from India saw near Ghorian, west of Herat, "some windmills for grinding corn; they are constructed on the same principles as those of Europe, but instead of canvas wings, broad-leaved flags are substituted" (Forster 1808, 156). His "flags" are presumably rushes. Forster's unquestioning acceptance of horizontal mills may be due to a brief vogue for them in England at the end of the eighteenth century (Wailes 1968, 135). By contrast, in 1832 Mohan Lal, the Kashmiri interpreter for Alexander Burnes on his famous mission to Bokhara, saw windmills for the first time, at Khaf; one of his European employers told him that "their clumsy and rude machinery was not to be compared with those of civilised countries. These mills go horizontally, and in other countries they sail perpendicularly" (Lal 1834, 184). It was midwinter, so they had no chance to discover that the "rude machinery" was both efficient and better suited to Persian needs than the sophisticated European windmill. The point was made again in the late 1960's when a wooden European-style mill was installed in one Khorasani village; after its second season they got rid of it because the "civilised" machinery needed too much maintenance and they did not like the quality of flour it produced.

A later traveller, H.W. Bellew, was one of the very few Europeans to mention the use of these windmills for water-raising. His observation in 1872 was only at second-hand, but it reminds us of the intriguing question of the non-application of the principle of gearing to Persian corn-grinding windmills (see below). The use of windmills for irrigation took the fancy of Chinese visitors to Khorasan in 1414. Such windmills are still used for this purpose in parts of China, but they preferred to continue grinding corn by water-power. In describing the ruins of Sistan early in this century, C.P. Tate makes the sole reference to another possible purpose of the Persian windmill, erected on the walls of "manor-houses ... to grind wheat and also to act as a thermantidote to ventilate and cool the inner apartments" (Tate 1910, 251). He was a careful observer, but one is tempted to imagine he has here confused, in the jumble of weathered mud brick, windmill with wind-catcher (see Chapter V).

Twentieth century practice

In the last hundred years certainly Persian windmills have been almost exclusively used for grinding corn and occasionally pulse (*nakhod*), while qanats and animal-powered saqiyahs have provided irrigation. Complex applications of wind-power like the sawmills and paper-mills of Holland have not been attempted (Reynolds 1970, 174f.). The windy season coincides with the end of the harvest and a shortage of water for driving water-mills, so the windmills have concentrated on the production of flour.

From explorers like Sven Hedin, consular officials like Percy Sykes and travellers like A.H. Savage-Landor a list of villages containing windmills c. 1900 can be put together. However incomplete this list may be it confirms the general geographical distribution already noted and it acts as a starting point for attempting an up-to-date survey. In the 1930's the Austrian geologist and explorer Alfons Gabriel, made several long journeys in eastern Iran. His references to windmills operating at the northern end of the 'corridor' are the last worth studying before post-war decay set in. Hans Wulff based his careful technical description on the windmills of Neh, but he included no estimate of the range of the Persian windmill (Wulff 1966, 286f).

If there appear to be few windmills today, one reason is that few people live in the region between the main road from Mashhad to Zahedan and that from Herat to Kandahar. Much of the area is desert and not readily accessible. The background to this situation has been examined in Chapter I. The other main reason is that the oil-based prosperity of modern Iran has brought cheap and easy power to most formerly remote villages: electricity and diesel oil grind corn at the flick of a switch. The work is fast, comparatively simple and independent of the weather. One man, working all the year round, can now supply the needs of a small village where perhaps a dozen peasant farmers used to spend much of the late summer operating their windmills. Windmills have fallen into total disuse in Sistan and the famous row at Neh has not worked for so long that local people now remember little about its active days. The fifty windmills of Khaf have been replaced by five diesel-powered mills and in nearby villages like Birrabad and Khargird a single windmill revolves occasionally because its owner prefers his home-ground flour. When the owner dies or a vital part breaks, the mill will fall silent (*khabideh*) like its neighbours. Only in Nishtafun (see Pp. 98-101) were many windmills operating, keeping alive Gabriel's impression (of nearby Sangan) in 1933: 'one hears the grinding of the numerous wooden windshafts from all sides of the village' (Gabriel 1935, 165).

The much greater poverty and far slower pace of development in Afghanistan lead one to expect that windmills would still be flourishing there. But the fine windmills of Ghorian, carefully recorded by Klaus Ferdinand in what looked like their heyday in 1960, had been so completely abandoned by 1977 that even the old millstones had been dragged off to line drainage ditches (fig 118).

Deliberately simple technology

The fact that the same fate has befallen the far more complex windmills of Britain, where roller mills had largely superseded them by 1900, may cause us to consider briefly why no advantage of the principles of gearing was ever taken in Persian horizontal mills. It would not have saved them in 1970, but would it have served them well before the oil-rich years arrived?

The geared Vitruvian watermill is found on the Plateau, albeit only along the Zayandeh Rud (see P. 84) and a knowledge of gearing is evident in the much more widespread water-raising device, the *saqiyah*. Why did the horizontal windmill stick to direct transmission? Certain observations help in formulating an answer:

a. The complication of gears would necessitate a much greater degree of supervision and maintenance than the simple horizontal mill requires.

118. A fine row of mills at Ghorian, working when K. Ferdinand recorded them in 1960, but utterly derelict in 1977.

b. Without the facility of gearing up the speed initially transmitted from the power source the grinding process is slow.

c. The wind-wheel turns in lighter winds than would be possible if the shaft was connected to gearwheels whose initial intertia must be overcome.

d. Only one pair of stones can be worked in each mill.

These circumstances fit the requirements of the peasant owners of these mills, or have possibly moulded them into the pattern which follows:

i. Essentially each miller is a farmer who spends some of his time in the summer months grinding his own corn for his family's flour supply. As was formerly the case in the Shetlands with watermills, the mill is almost part of his farm equipment.

ii. He has little technical expertise in engineering and carpentry. Therefore both maintenance and rebuilding after an earthquake must be straightforward.

iii. He wants a slow, simple system which does not require his attention too frequently. There is not enough grinding to employ a man full-time. It suits the farmer to run the mill for a few hours some days and to leave it with a full hopper for much longer on other days, while he is busy in the fields.

Traditional mistrust, harsh landlords, poor communications, an absence of manorial milling rights (the medieval European *soke*) have also curbed any incentive to become a full-time miller. Each farmer has his corn ground in the family mill if possible, perhaps under the supervision of a member chosen for the task because old age or illness have made him unfitted for work in the fields. The extended family and the patterns of inheritance mean that many men may own small shares in a windmill; thus all their interests may be best served if one of their number handles their mill during the four grinding months. He is however no skilled craftsman and figure of local standing as were the men who operated the large vertical windmills of Europe.

Ease of operation

There are few tricky, precise operations required at a moment's notice in these simple horizontal mills. No part of the mill has to be turned into the wind. There is no fear of tail-winding, as occurs when a sudden change in direction of the wind can badly damage the cap and sails of a vertical windmill. The single hopper beside the stones on the ground-floor is filled, as necessary, from the nearest convenient sack. No other machinery within the millhouse needs supervision. The only alteration the miller can make to his sails is to decrease the force of the wind acting upon them by placing across the wind inlet a reed screen (fig 119). There are only two bearings requiring lubrication: only the lower one, the lightening cone (*mosht*), must be removed for this purpose. It is a stout wooden spike, c.40cm (1ft 4ins) high, with an 8 cm (3ins) diameter base and a blunt, rounded tip, resting on a massive foot-beam. It is remarkable in that it alone supports the entire weight of the wind-wheel as it presses down on the iron rynd set into the lower face of the runner stone (fig 109f). The cone is greased with mutton fat and must be replaced on average every three weeks when a mill is working more or less continuously.

119. Reeds across wind slits on a gusty day. Reeds and sacking lessen the force of the wind through the wind slits at Nishtafun, 1977

In fact such occasions occur at irregular intervals since not only does the 120-day wind blow with very varying intensity, but also when it gusts fiercely the windmills have to be brought to a standstill. With the direct transmission of a horizontal mill the stones and wind-wheel revolve in unison; at a fast speed the ram-shackle wind-wheel threatens to disintegrate and the flour emerges from the stones in the same warm, slightly scorched condition that traditionalists hold against the product of the diesel-powered stones which have now replaced most of the Persian windmills. A wind speed of 25 mph produces c.25 rpm of the windshaft and of the runner stone and is considered by Persian windmillers to represent ideal conditions for their corn grinding. Geared English windmills drive stones at four or five times faster rpm, though stones are seldom as wide in diameter as the standard 150cms (5ft) of Persian stones.

There is no carefully dressed geometrical pattern of furrows and lands, only a roughened surface at eye and skirt of the stones (see P. 100). The stones, quarried near Nishtafun, weigh roughly one ton when new. This is the weight which the lightening cone must bear and balance accurately on its rounded top a fraction of a millimetre above the bed-stone. There is no delicate tentering mechanism as in English windmills: no screws, no steelyard, no governor. Nor in any Persian mills was lead seen added to special hollows in the top surface of the the runner stone to ensure absolutely correct balancing, as is the practice in English mills. When the consistency of the flour is unsatisfactory the miller marginally alters the height of the lightening cone, by knocking in or easing out wooden wedges inserted beneath the footbeam, so that the gap between the stones may be corrected (Ferdinand 1963, 83 illus.). This elementary 'rule of thumb' technology, which demands no specialist skill, clearly suits Persian traditions of casual milling.

Working windmills

In 1977 the village of Nishtafun, south-east of Khaf, retained a large number of working windmills and a pride in their possession and performance which was hard to find elsewhere. Thirty-eight mills were grouped in five separate rows along the northern edge of the village (fig 120). Most were still being used by the families who owned them, grinding wheat for themselves and their friends (who paid a 5% toll) as often as the wind permitted and the state of the domestic flour bin dictated. The village is not as self-sufficient in corn as it is in milling facilities, since the water supply is less easily exploited for irrigation than on the plain at Khaf. There were 1891 inhabitants in 1956, before any diesel-powered mills were installed. In 1977 there were c. 3,000 and there was still plenty of business for the mills, whose operators seemd prosperous and confident. They would quote no agreed figure for average amounts ground per hour, day or season, since all depends upon the wind and the custom they receive. A rough average, derived from discussions in windmills in various villages, is 300kg (660lbs) in any twenty-four hour period. A dozen mills working continuously during th windy season could theoretically have supplied Nishtafun's flour needs in 1956. In practice, of course, each mill only ground to fulfil the demands placed on it, that have been outlined above.

The mills here face N.N.E. and their position on a bluff above the village, with only the graveyard between them and the open plain, is so favourably exposed that certain other hamlets bring grain here for ginding. Until recently the slow pace of camels and donkeys, the problem of finding them fodder and water, and the threat of marauding Turkmen and Baluch tribesmen in this frontier zone discouraged much traffic of this sort. Now lorries have extended the potential range of operation. However, the same fuel that powers the lorries can also drive a diesel mill in any village.

The millhouse

A typical mill at Nishtafun is 10m (32ft 6ins) long and 5m (16ft 3ins) wide, built with three bays: in the

120. A row of working mills at Nishtafun, viewed from the south, 1977.

windward bay, raised a metre or so above the floor, are the stones and grain hopper; in the middle bay flour is stored while awaiting collection; in the bay nearest to the door grain is winnowed. The grain hopper is made of mud and fixed to the wall rather like a huge swallow's nest under the eaves on the western side of the stones. Flour swirls out on the opposite side, which is paved with stone slabs. Here the miller sweeps it up and weighs it before filling the customer's sack (fig 109e).

Stone dressing

In this windward bay too, the dressing of the stones takes place. The miller uses no special tackle to raise the runner stone, but he is dependent on a number of strong and willing neighbours. They bear down on a long pole, called *damardta* (the strength of ten men), inserted under the footbeam, so that the lightening cone is pushed upwards sufficiently to place wooden rollers, the size of cotton reels, between the millstones. The wind-wheel is raised slightly further on the roof and a long peg thrust right through a hole in the windshaft to bear its weight, resting on the millhouse roof (fig 121). Back within the mill the lower end of the shaft, the *peiwand*, which contains the thrust-pin – now freed from the rynd, is detached. This peiwand is rather over one metre long and

121. A windshaft raised and braced on the roof.

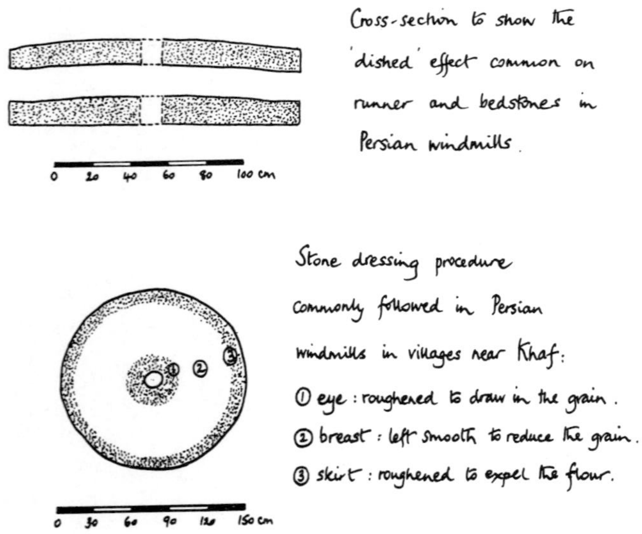

122. The piewand *or bottom section of the wind shaft in an Afghan mill. Notice how it is pegged and bound to the shaft; also the suspension of the feeder channel from the grain hopper and the vibrator stick responding to the uneven upper surface of the revolving runner stone. (photo R. Hewer, 1969). (left).*
123. Persian millstones and their dressing. (above).

is attached to the windshaft by wedges and ropes (fig 122). Once it has been removed it is possible to slide the runner stone off the bedstone and manhandle it upright against the side wall. The miller then pecks the surface of the eye (centre) and skirt (edge) of both stones with a steel bill to ensure the right breaking open and grinding of the grain. The middle section, between eye and skirt, known as the breast in English windmill terminology. is kept smooth. At the eye the grain is cracked open, at the breast it is rendered small, at the skirt it is ground into soft, fine flour. The surfaces of the stones are very slightly dished and the 10 cm (4 ins) wide skirt is flattened – like a shallow inverted soup plate (fig 123).

Stone quarrying

One reason for the survival of the milling tradition in Nishtafun may be because the millstones are cut from Kuh Seeno, a mountain only six miles away. This formerly supplied the needs of all the windmills of eastern Khorasan. Anyone could quarry his own stones, but since this represented a fortnight's work and was only necessary once or twice in any miller's lifetime, mill-owners from distant villages like Khaf or Niazabad preferred to purchase them from the men of Nishtafun, who thus alone acquired real expertise in this laborious activity.

Threaded on a stout pole to which oxen were yoke the stones were then trundled to their destination, exact as for watermills (see P. 80). A team of men was needed to steady the ends of the pole and generally ease its path on the slow journey which might occupy an entire day. On arrival the stones would need some further shaping and five days of running-in, during which sand, then millet and finally barley were gradually fed through. This operation is uncommon now, since there is greatly diminished dependence on the windmills for grinding; and unwanted, partly used millstones are often available in mills which have ceased working. A pair of stones is reckoned to last anything up to fifty years before the runner has become too thin and light to grind properly. In 1956 a new pair cost the equivalent of £17.

Grinding

The survival of the windmills of Nishtafun is in par due to the dislike some villagers have of the warm flour produced by the diesel mills with their small, fast-runni stones. They maintain that the soft, cool flour, slowly

124. Pegs laid in the feeder channel of a mill at Birrabad. Notice also the mud collar raised round the eye of the runner stone and the small wedges keeping the iron band in place round the foot of the windshaft (1977). (left).
125. A windwheel at Nishtafun braked by two forked sticks, 1977. (right).

ground by windpower, has greater nutritional value. In the gusty conditions of the windy season a slow feed of grain is essential, otherwise the stones will become hopelessly clogged up. Since the miller is often away on other business he ensures that this will not happen by laying a number of pegs in the feeder channel. These pegs, resembling short, stubby pencils, attached by strings to the grain hopper, cause the grain to dribble into the eye of the runner stone very gradually (fig 124). The feeder channel is joggled by a vibrator stick attached to it at one end while its lower end bumps over the uneven upper surface of the revolving runner stone (fig 122). It is named after the stork in Persian (*laklak*), perhaps because the sound resembles the bird's clacking beak (c.f. the terms "clacker" or "click mills" for the horizontal watermills of the Orkneys and Shetlands). The stone moves slowly and erratically in response to the fluctuations of the wind. The stronger the wind and the higher the rpm of the runner stone the more grain is enabled to pass through the obstacle course presented by the pegs in the feeder channel. So the quality of the flour should remain fairly consistent.

A wind speed of c. 25 mph (i.e. 'Strong Breeze') has been quoted as ideal (see P. 98). The wind-wheels will keep turning in lighter winds, down to c. 8 mph. Once the wind is regularly gusting well above 30 mph the miller has to close down his mill, since the reeds rising half the height of the wind-slit, even when covered with a rug also, cannot sufficiently moderate the force of the wind. The light framework of the wind-wheel is braked by inserting forked sticks resting against the side walls (fig 125) and the lightening cone can be lowered so that the runner stone, together with the windshaft, rests on the bedstone. 'Whole Gale' conditions are experienced roughly once a month in the summer. Then not only the millers batten down their hatches; all men retreat indoors from the fierce blast which swirls up the desert dust as it passes. Such strong winds (c. 60 mph) may last for 24 hours. More often the light breeze at sunrise increases steadily to an unwelcome gale by noon which then subsides around sunset. So the mills are busiest in the morning.

Time for reflection

Once the hopper is full the miller will squat beside the open inner door of the winnowing tunnel and prepare the next load. The corn will have been winnowed after harvest in the fields; now he goes carefully over it using a mallet to knock off any recalcitrant husks and a large sheep's gut sieve. He sets aside the cleaned grain on a cloth. 2% of rock salt is added in due course and the grain is stored in one of the mud-walled bins along the side of the millhouse until the hopper runs low.

A neighbour may call to ask for help in raising his wind-wheel or runner stone, or a regular customer may arrive with a donkey load of wheat to be ground, or the miller may be able to sit for a while in the pleasant draught and the shade within the mill, out of the strong parching wind and the burning sun. These are quieter months for the peasant farmer: his harvest is in and the autumn ploughing lies in the future. However, his vegetable plot requires regular irrigation and the produce must be prepared for the next market-day. This is the building season too, for there is time to add further courses to mud brick vaults and to protective garden walls left unfinished months earlier. While the wind from the northern deserts spins the windmills the inhabitants of Nishtafun draw breath and take stock before the winter comes.

Any future for these mills?

It seems very probable that all the hoppers and bins at Nishtafun will soon run dry, as they have in the other windmill villages of Khorasan and Sistan. Diesel-powered grinding, which has ousted traditional methods elsewhere, has failed so far to dislodge the Nishtafun windmills, but in July 1977 that other twentieth century wonder-worker, electricity, was literally on the doorstep. The poles were up and huge drums of cable were slowly unwinding from the main road. Traditional technology of small unsophisticated communities on the Plateau is unlikely to hold its own. Even though mud brick buildings will continue to be favoured – for good reasons, as Chapter II indicated – corn-grinding will be carried out in them faster and with less physical effort without harnessing the wind. Millhouses will be used for storage, while the wind-wheels gradually disintegrate. Their owners will cease to regret their passing, like the men of Neh who ignore the derelict row of sixty windmills (fig 111) which bounds their village on the north beside the new tarmac road. Only the occasional enthusiast will survive, like the old man at Khargird whose mill had been listed and recorded in photographs by the Historic Monuments Department at Mashhad. He was keeping his mill in excellent order in 1977, doing some infrequent grinding for a few friends who preferred his flour, and lamenting the fact that his son, a Post Office employee, would have no interest in preserving this heritage when his father died. One hopes that the authorities in Mashhad will remember their commitment, so that at least one example of a small but significant Persian invention may survive to illustrate the history of technology on the Iranian Plateau (fig 126).

126. A working mill and its proud owner in 1977 (left).
127. Mills in the great row at Neh, now derelict (photo T. F. Hewer, 1971). (right).

8 PIGEON TOWERS

128. Pigeon towers outside Isfahan (1970).

Pigeon towers are not typical of the Plateau but they have long been landmarks in the Isfahan oasis. "I don't think there are finer Dove-cots in any part of the World" wrote Sir John Chardin in the seventeenth century (Chardin 1927, 177). Even in ruin they are spectacular and hundreds still dot the hazy green sea of orchards and gardens which surround the city. Massive in scale, these structures seem to be incongruously out of context; they might be Martello towers stranded hundreds of miles inland, or chess-men awaiting the master mind of a remote giant.

The bigger towers are free-standing but many of the smaller, built into the walls of the gardens, are deceptively akin to bastions or corner towers in a defence system. Others brood protectively, but unstrategically, over the flat mud roofs of village houses. Their useful but unromantic purpose is to collect the pigeon manure which has been found to be so beneficial to the melon fields, but the sculptural form and the fascination of pattern of the interiors of these utilitarian structures is astonishing.

As in all traditional building dating is difficult. The only two (figs. 137-8) to which even a period is ascribed are thought to have been built during the reign of Shah Abbas (1587-1629) in the great royal gardens of the Hazar Jarib ("thousand acres"), but since these have more highly developed plans than any others now extant it must be assumed that a considerable tradition lies behind them.

Unfortunately, there is a dearth of travellers' reports in the period preceding that of Shah Abbas when they

129. (Top) A tower in use in 1961; the buttresses are a recent addition. The bold bands of red ochre and lime wash give the tower a nautical air. Long wooden water spouts can be seen projecting from the upper turret and the main tower.
(Left) A small tower at the corner of a garden wall (1963).
(Right) A badly cracked tower, typical of many (1963).

appear, architecturally, in full flower. They were first noted by the seventeenth-century traveller, Thomas Herbert (1628) between Shahreza and Isfahan (in Mahyar): " ... albeit their houses were neat, yet they were in no wise comparable to their dove-houses for curious outsides" (Herbert, 1928, 120). For the next three hundred years traditional designs seem to have been handed down within each family or village but although the objective of the builders and the unit on which they worked, the pigeon hole, were identical, no towers seen were the same.

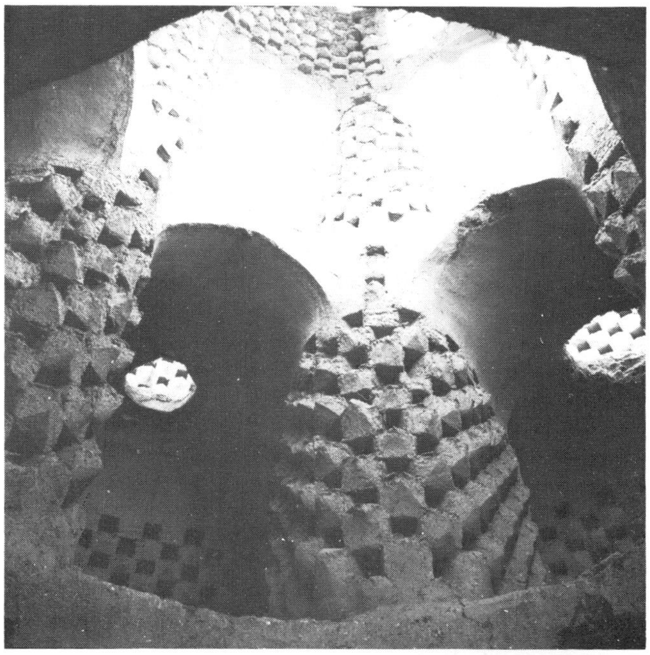

Single round towers

Amazing inventiveness has gone into the solution of the basic problem: the provision of the maximum number of pigeon holes with the minimum amount of building material. That material being unbaked mudbrick, plastered with mud, this requires ingenuity. Timber is rarely used,

130. Interiors of ruined towers. Pigeon holes and perches cover every available vertical surface (1963).

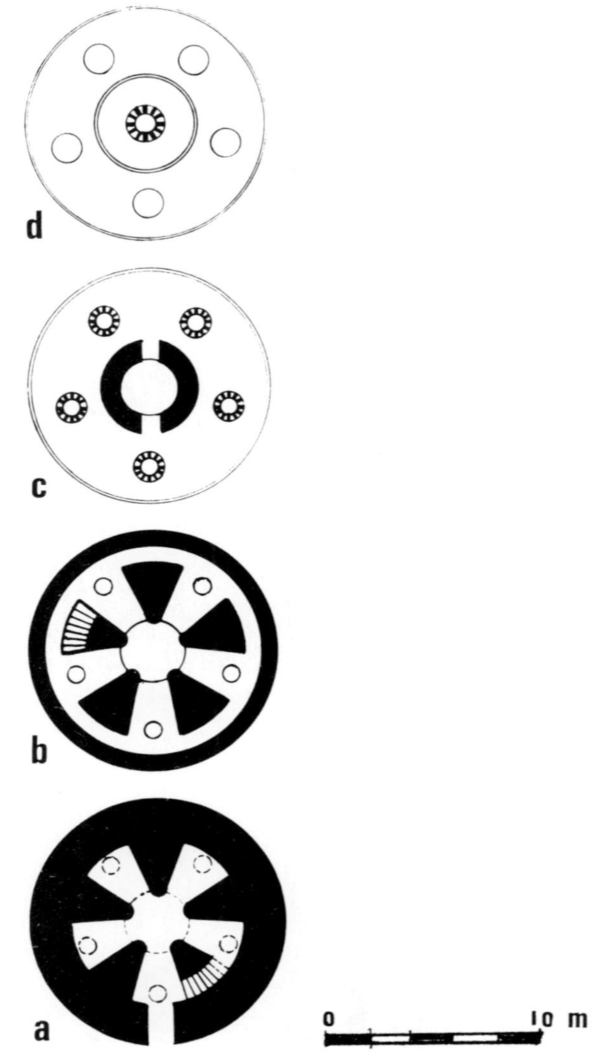

131. Plans of a "typical" tower: (a) the ground floor; 5 giant buttresses, like the spokes of a wheel, support the drum of the central turret. Steps lead to the upper galleries. (b) the first and second floors; the galleries are pierced by holes which allow the birds to fly down from the pepper-pots on the roof, through which they enter the tower. (c) the main roof is topped by five pepper-pots. (d) the upper roof; one pepper-pot crowns the central turret.

132. A section of a "typical" tower.

133. A "typical" tower. Looking up to the pepper-pot which crowns the central drum (1963).

so the whole structure must be designed to be in compression, minimimising tensile stress. Circular ground plans, which are by far the most common, are important in this respect. The variety of plans within the basic concept of a circle, shows great imagination and, with their curving vaults and pierced saucer domes, produce fascinating sequences of solid and void (figs. 130, 133). Had Herbert seen the interiors he would have been even more amazed.

Basically the towers consist of an outer drum, battered for stability and buttressed internally to prevent collapse and to gain lateral support from an inner drum which rises perhaps half as high again as the main structure. The main drum is divided vertically by the galleries which cut the buttresses and are connected by a circular stair. The galleries are supported on barrel vaults and saucer domes. Between the buttresses (which look like the spokes of a wheel on plan) the domes are pierced to allow the birds to fly up and down; similarly the inner and outer drums are connected by open arches at every level. The pigeons enter only through the sides of the domed cupolas or "pepper-pots" of honeycomb brickwork. One of these crowns the inner drum while others usually ring the flat roof of the main drum below. They vary in number according to the ground plan; half a

134. The huge tower at Chahar Burj (1963).

dozen would be usual but one tower (still in use in 1961, at Chahar Burj) had twenty, plus four on its central drum (see figs. 134-5). A tower in the Hazar Jarib, recorded in the early eighteenth century, had a second tier of pepper-pots crowned by those on the central drum (Kaempfer 1712, Pl V).

Size varies considerably. A big tower might have a base diameter of 12 metres (c.39 feet) and an overall height of about 20 metres (c.66 feet). The Chahar Burj – the biggest seen – was 22 metres (c.72 feet) across.

Most builders seem to have been content to build the outer wall as a simple drum; it is the way in which it may be alternatively hollowed out and buttressed internally which provides its architectural fascination. However, the two towers attributed to the late sixteenth or early seventeenth century in the Hazar Jarib (figs. 137-139) have the further refinement of a corrugated outer wall which increases the stability of these large towers without increase of wall thickness. This also increases the surface area of the wall and hence the number of pigeon-holes. The eastern tower could be thought of as a cluster of eight small drums round a larger central drum.

The mesmeric character of the inside of the towers comes from the repetition on every vertical surface, wheth-

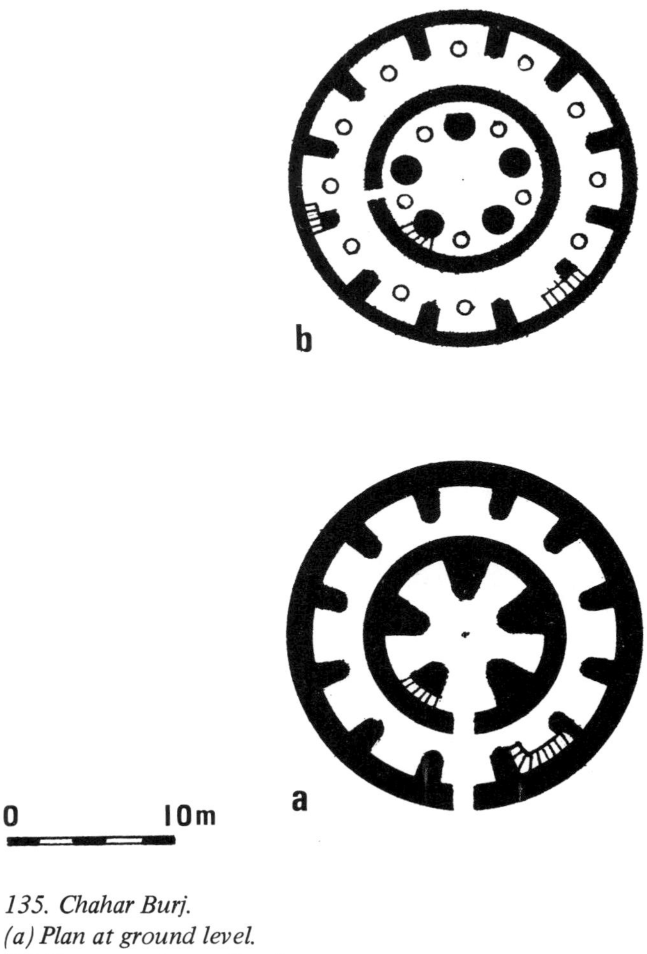

135. Chahar Burj.
(a) Plan at ground level.
(b) Plan at main roof level.

136. The interior of the eastern tower of the Hazar Jarib: the central drum (1963).

er or not it is curved on plan, of the standard pigeon-hole, 20 x 20 cm by 27 cm deep (c.8 x 8 x 11 inches) with its mud perch below (fig. 130). Much of the sculptural quality of the structure is due to these perches; note the comparative flatness of those interiors where they have fallen away. Each is made of an asymmetrical mud pyramid of four unequal sides whose square base was clapped, damp, on to the vertical brick face below the hole. Whether this was done at the time when the brick was made or when it was in position is not clear. It would seem easier when the brick was on the ground and still 'green', but it would have been awkward to handle. When in position, the smaller top side of the pyramid forms a horizontal perch and the other sides slope away, making access to the neighbouring holes easier for the pigeons next door. This perch allows the bird to alight clear of the nest and droppings to fall clear to the floor below – an important point since the function of the tower was to collect manure.

137. The western of the two surviving towers of the Hazar Jarib, believed to date from the reign of Shah Abbas (1586-1628). (Photo 1963).

138. The eastern tower of the Hazar Jarib (1963).

139. Plans of the towers of the Hazar Jarib.

140. Painted towers near Lenjan, 1971 (see also fig. 142). Left, a cluster of four topped by a central turret. Right, a battery of sixteen. There was an ice-pool in its shade.

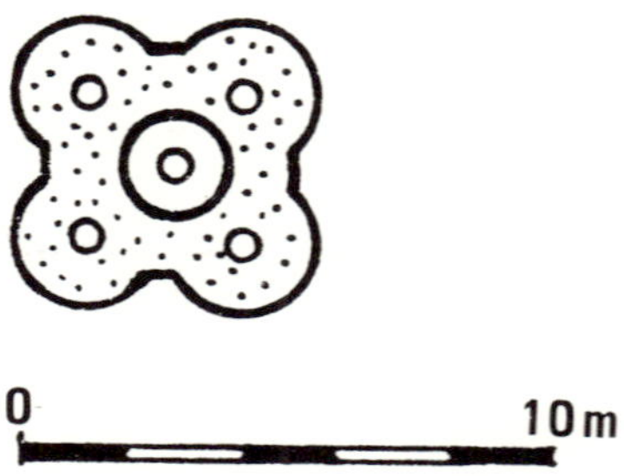

Clustered towers

Although the vast majority of towers are single drums two very unusual designs in which drums adjoine were seen in the village of Kelisan near Lenjan (1970). One consisted of a cluster of four small turrets of about 3 metres (c. 10 feet) in diameter crowned by a central drum (fig. 140).

The other, built on the edge of an orchard, consiste of a battery of linked turrets which was 38 metres (c. 12 feet) long, each turret having a diameter of about 2.3 metres. The last two turrets at each end had two more attached, making a cluster like the one already described (see plan below). These projections gave shade to the ends of an ice trench that had been excavated in the lee of the wall (see Chapter 4, P. 51). A battery consisting of two adjacent rows of five towers surmounted by four more was photographed near Pir Bakran in 1976; they too were painted. They were not in use.

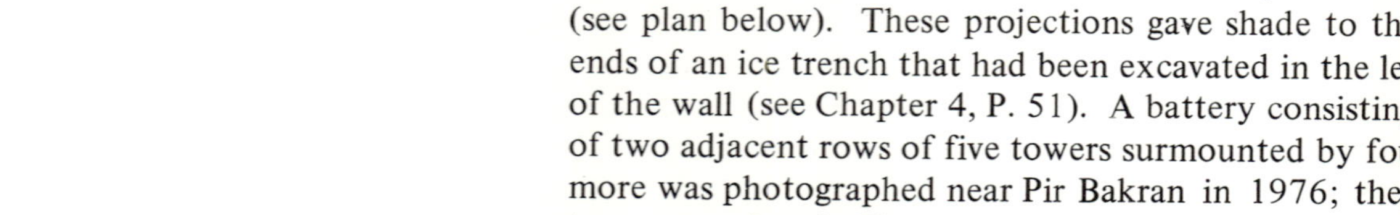

141. A rectangular pigeon tower near Gulpayegan, 1971. Pigeon holes pattern the face of the tower. The plan below is drawn to the same scale as that of the cluster opposite.

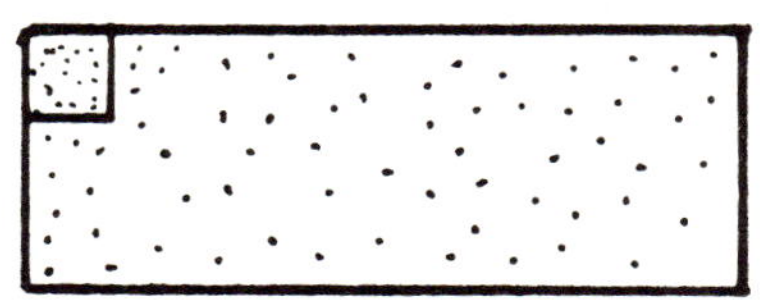

Rectangular towers

Big rectangular pigeon towers of quite different design were found in use (1971) in the Khunsar-Gulpayegan region, 125–200 kilometres to the northwest of Isfahan (see fig. 141). A dozen were seen near Khunsar; there were many more around Gulpayegan. They were reminiscent of small forts. Some were sited singly; others were grouped; none seen were built into garden walls. A typical tower measured 12.3 x 4.5 m (c. 40 x 15 feet) at the base and was 7 - 8 m (c. 23 - 26 feet) high. The walls, which were battered at the base, were decorated by a broad plastered band 2 - 3 m (c. 7 - 10 feet) above the ground. This smooth surface was probably introduced to prevent the entry of reptiles (see below). The top of the walls was crenellated. It was not possible to enter any tower to make a plan, for the cheering reason that they were in use. The only square tower seen was in the ruined village of Jozdan, towards Gavart to the east of Isfahan. This consisted of a two-storey outer structure, 10 m (c. 33 feet) high, surrounding a three-storey inner tower. Surprisingly, there was also a large number of small pigeon towers (of the usual circular plan) set close together in the same village.

142. A detail of the decoration on one of the Lenjan towers.

Snakes and decoration

The towers are entered once a year for the collection of manure, after which the small doorway (occasionally there are two) is bricked up for the next twelve months. One tower (presumed to be in use as it was in very good repair) had no entrance below roof level. High level entrances are not uncommon; they are almost certainly positioned to reduce the danger of snakes and rats getting into the tower. The cause of structural cracks (fig 129)) was said to be the tremendous vibration set up by the wings of hundreds of terrified birds if a snake got in. Some cracks may have been caused by earthquakes, but most of those in the many disused towers still standing were probably caused by rain. Deterioration would be swift once the long wooden water spouts which take the water clear of the battered walls had gone; they are seldom left in an abandoned tower.

External decoration varies according to the grandness of the tower and the whim of the owner but even at its most exotic it probably derives from the dual function of keeping snakes and rats out and helping the birds to get in.

A band of smooth plaster, about a metre wide, often coloured in lime-wash or red ochre, decorates the main wall (fig 129). It would prevent a snake getting a grip; it might otherwise creep up the battered drum of the tower on the rough mud and straw plaster with which it is rendered. Projecting courses of brick and plastered cornices and friezes, besides giving an effective decorative capping to the wall, also provide projections which snakes would find difficult to negotiate. Painting is now relatively rare but Morier (1810 - 16) noted that towers were "painted and ornamented" and that "more care appears to have been bestowed upon their outside than upon that of the generality of dwelling-houses" (Morier, 1818, 140f). The cornice of even the simpler towers usually has a "spoon" pattern moulded in the plaster; this seems to be the remains of the idea of finishing the tops with stalactite decoration which was so popular among Muslim builders. In a late nineteenth century engraving of a tower, a double row of stalactites surmount a painted frieze but unfortunately it is difficult to see what the paintings represent (Dieulafoy 1887, 285). About thirty years before towers "painted with all manner of strange devices" had been recorded (Binning 1857, 124f).

The only modern example seen of this tradition of painting survived on the groups of towers already described (Lenjan) to the south-west of Isfahan. Here the tra-

143. Towers illustrated in Morier's Second Journey *(1810-16).*
144. Brick detailing at the top of a tower is functional as well as decorative. The smooth plaster band has almost disappeared (Herb Garcia, 1963). (top right). Turrets through which pigeons enter a tower (bottom right).

dition was very much alive (figs 140, 142). Astonishingly vigorous decoration in cobalt blue, black and red ochre on white depicts stylised birds, pomegranates on twining branches and, surprisingly, a man (the owner?) in a big black hat. The battery of towers nearby was decorated in a similar way but here the head of a woman appears and she is unveiled which further suggests that the towers were owned by an Armenian. However, immediately below the cornice is a painted frieze of decorative script giving the Islamic date, 1335 (1952 A.D.); a brick inscription on the battery could be read as 1324 (1946 A.D.).

The unique grouping of these towers as well as their decoration, suggests that the owner, painter and builder – perhaps the same individual – were exceptionally inventive; certainly the painter had a nice sense of humour. His designs also typify the traditional Persian enjoyment and masterly management of pattern and colour; the painted bands whic'' are densely filled with elaborate decoration are boldly contrasted with areas of blank wall space.

Strikingly painted pigeon towers have also been recorded relatively recently in Afghanistan in the Hari Rud valley between Herat and the Iranian frontier. Square, rectangular, round and even an octagonally topped tower were described and illustrated nearly forty years ago (Zestovsky 1949, 17ff). The square tower was decorated by two huge cats which confront each other below a broad freize painted with zigzag lines; it was claimed that leopards are known to protect pigeons. Two more such cats decorated a rectangular tower seen in 1977 at Ghorian; this design was in brickwork, in low relief. Zestovsky also mentions that in these towers, which were 14 - 15 metres (c. 48 feet) high, the first floor was used for food storage and the ground floor for beasts. Pigeons entered the upper part of these towers through honeycomb holes in the walls of the tower itself; these holes give a contrasting texture to that of the smooth wall surface.

In the Isfahan towers the honeycomb brickwork which gives the pigeons access through the little cupolaed turrets, gives the turret walls a surface pattern which contrasts nicely with that of the main tower. This brickwork is sometimes carried round the tower as a balustrade giving the birds more places where they can perch.

Melons and pigeon dung

The melons of Isfahan were noted as early as the fourteenth century when Ibn Battuta rated them third, coming after those of Bokhara and Khwarizm in gastronomic excellence (Gibb 1971, 547). It by no means follows that the melon fields of the middle ages were fertilized by dung collected in pigeon towers although this has long been prized for its value to the melon crop (as it is now for the small cucumbers which are an important part of the Persian diet).

Today, as in the past, the only function of the tower is the collection of this manure. In a country where most animal manure is used for fuel, pigeon dung is of particular value for the land. It is as renowned in Isfahan as it once was here, being exceptionally rich in phosphorous and nitrogen; it is mixed with ash and soil in varying proportions depending on its use: "They call it *Tehalgous, i.e.* enlivening" (Chardin 1927, 177). Certainly the only way to collect any quantity of such dung is in a pigeon house where it gradually dries out and is easy to cart away. Slight saltiness in the soil in which the melons are grown is also thought to improve the flavour of the fruit,making them more refreshing in hot weather.

The great effort required to build the towers must be matched against the demand for dung to provide enough melons, "a fruit that is indispensable to the existence of the natives during the great heats of Summer" (Morier, 1818, 141). It was the shift of the capital from Isfahan to Tehran, causing a decline in the market for melons, which resulted in large numbers of pigeon towers becoming redundant early in the nineteenth century. It is hard to say how many there were in their hey-day. Chardin writes "They reckon above three thousand ..." (Chardin 1927, 177) and someone may have known since a tax was paid to the Shah on dung. Any Mohammedan could build a tower, but not a Christian. The large Armenian colony which had been settled in Julfa to build the new Isfahan for Shah Abbas may have found this irksome.

The only count of pigeon holes recorded was made about a hundred years ago. A "large" tower was said to have 7,100 pigeon holes (5,500 in the outer wall, 1,600 in the inner buttressing (Wills 1891, 130). It was claimed that the big tower already mentioned, at Chahar Burj, had space for 10,000 birds: a good round number. It was still in use on a very much diminished scale in 1961, but had once provided the chief income for the village. In the early 1960's about a thousand pigeons were thought to live and breed in the average tower and to yield about 6,000 pounds of dung in a year. The peasants claimed an increase of yield of at least fifty percent if the dung was used at a rate of "about 1,500 pounds per acre for most crops and 2 pounds per fruit tree" (Wulff 1966, 270).

When whole villages belonged to one land-owner who also owned the pigeon-towers the pigeons naturally fed at the expense of the tenants as well as on the landlord's crops. But since the important crops around Isfahan in the hey-day of the pigeon tower were melons and other fruit rather than grain, the birds must have foraged a good deal on uncultivated land or fed further afield.

Gunpowder and dung

Fryer (1672-81), on a journey up to Isfahan from Shiraz "encountered almost in every village old Castles made of Mud and almost turned to Earth again: in whose stead, at the Emperor's Charge, are maintained many Dovecots, pleasantly seated in Gardens, for the sake of their Dung, to supply the Magazines with Saltpetre for making Gunpowder, they have none else but what is Foreign" (Fryer 1698, 259). Curzon and others are surprised by this statement (Curzon 1892, 19). However saltpetre could be produced artificially by processing a mixture of earth, lime, ashes and pigeon dung. In England,pigeon-houses were described in 1604 as being "the chiefest nurses of saltpetre of the kingdom". At this time the powder contractors were to confine their operations to half-an-hour a day, and to compensate the owners for any pairs of pigeons or eggs lost. In 1625 a proclamation issued by Charles I forbade owners to pave pigeon houses with stone or brick or to floor them with anything other than "good, mellow earth" (Suffield-Jones 1966, 176). The earth flooring, typical of Persia, would allow for good drainage of the dung.

This evidence of English practice suggests that Fryer had his facts right and that pigeon dung was indeed of great importance to the State for the manufacture of gun-powder, anyway at the end of the seventeenth century.

Pigeons as food

Despite the high value of the manure it is surprising to an outsider that the towers were never used to provide pigeon meat, which has been their chief function in Europe. In medieval England, when the peasant had little redress if the landlord's pigeons ate his corn, Church as well as lay landlords tucked into pigeon-pie. In hungry Persia similar eating habits might have been expected.

The fact that we refer to these birds as pigeons and not as doves is in itself a reflection of the very long-standing tradition of sacredness surrounding doves. This goes back much further than the Christian association of the dove with the third person of the Trinity, having Sanskrit parallels. This free but semi-domesticated bird living close to human dwellings has often been felt to be an appropriate symbol for the soul. Its sudden appearance can be taken as an ill omen forboding death, or alternatively the killing of the bird may be regarded as an act of sacrilege.

Early travellers were surprised at this abstention from pigeon flesh on the part of the Persians. Marco Polo (who mentions no towers) wrote that in Rudbar (south Iran) "Turtle-doves flock here in multitudes ... The Saracens never eat them, because they hold them in abhorrence", (Marco Polo 1958, 33). It could be that the opposite reason was intended by his informant. But Morier wrote, "The Persians do not eat pigeons, although we found them well flavoured. It is remarkable that neither here nor in the South of Persia have I ever seen a white pigeon, which Herodotus remarks was a bird held in aversion by the ancient Persians" (Morier 1818, 141).

Probably Herbert gives us the clue when, after describing how much grander the pigeon towers of Mahyar were than the ordinary houses, he goes on: "This reason they give: some of them (as tradition persuades at least) are descended, not *a columba Noe* (i.e. from Noah's dove), but from those who, being taught to feed at Mahomet's ear, not a little advanced his reputation, persuading thereby the simple people they communicated to him intelligence from some angel" (Herbert 1928, 120). Incidentally, the 'simple people' of Wales had been similarly persuaded not so long before. The dove, often depicted on St. David's shoulder, also transmitted divine intelligence (which led to his refutation of the Pelagian heresy).

At all events, while Europeans lived well on pigeon squabs, the Iranian peasant, however hungry, seems to have abstained, except for those who kept a few *yahoo* (see below).

Pigeon shooting

Birds might be shot on the wing for sport but not if they had alighted on a tower. Olearius, writing in 1669, gives a curious description: "The King sent to us betimes in the Morning to invite us to go to a Pidgeon-hunting. We were carried to the top of a great Tower, within which there were about a thousand Nests. We were plac'd all without, having in our hands little Sticks forked at the ends. The King commanded our trumpets to sound the charge, and immediately there were driven out of the Tower or Pidgeon-house great numbers of Pidgeons, which were most of them kill'd by the King and those of his Company. This was the end of that kind of hunting" (Olearius 1669, 211). Perhaps it was not quite the end, for until recently it was a popular sport to drive the pigeons out of the qanats by throwing stones into the well and to catch the birds by hand or hit them with sticks.

Pigeon and pigeon fanciers

Several different kinds of pigeon are kept by the villagers. That of the pigeon-towers is the *kabutar*, the blue one, which is of the same family as the blue rock found on the west coasts of Scotland and Ireland. Carriers are rare: this may seem strange in a country where they might have been exceptionally useful but pigeons would have been unreliable messengers in a country where falconry was a popular sport and birds of prey abound. (Peregrine falcons in the U.K. have kept well-stocked larders without great effort by choosing nesting sites on cliffs below the regular routes of racing pigeons).

In some parts, villagers keep the fantail (*baba kooh*) which is believed to be that which perched on the Prophet's shoulder; devils are dispersed by the least flutter of its wings. The *yahoo*, the pigeon with feathered legs, is bred for its flesh and kept with the roosters; it also prevents devils harming the children of the household and their cooing, *yahoo*, gives protection to the house. They are also said to sense the coming of an earthquake and will fly out of a building a few seconds before the shock (Donaldson 1938, 163).

The tumbler (*mallagh*) was kept for sport. The sport of flying tumblers (*kafteh-bazi*) was popular in the nineteenth century, but, sad to say, it seems to have died out. The pigeons were flown morning and evening, perhaps about thirty from a loft. They would circle wider and wider above the village roof-tops until they were called back for feeding by their owner waving a stick with a coloured rag. The weaker birds would return first and the strongest last. As they approached home they would begin to tumble down in the sky, some birds making as many as twenty somersaults, to land on the owners head and shoulders. One popular pastime was to fly the flock close to that of a rival in the hope that they would 'capture' the outlying birds from the other flock and bring them back to their loft. Both owners would frantically encourage their own birds, calling and waving their sticks. At other times the fanciers would creep along the roof-tops to net another's tired bird using a fantail as a decoy.

Favourite birds were decorated with rings of silver, ivory or brass. These rings were not returned when the birds were claimed but were considered a lawful prize. Their value was such that some people seem to have lived off them.

The men who flew tumblers were known as *luti*, blackguards. This may have been partly on account of the prize rings they collected,but it was also accounted for by the fact that this pastime meant they spent much of their time on the roof-tops, probably looking into other people's courtyards and this was held to be the height of impropriety (Wills 1891, 94f).

Chardin has a seventeenth century version of the pigeon-fanciers' games "One of the best Sports the Rabble can have, is to catch Pidgeons in the Country, and in the Cities also, tho' 'tis forbidden. They take the wild Pidgeons by means of the tame ones, which are bred up for that purpose, and which they cause to fly all the day-long after the wild ones; and all those they meet with they take in amongst them in their own Flock, and carry them to the Dove-Coat.

Sometimes tame Pidgeons bring away likewise other tame Pidgeons from their Dove-Coats; so that all of a sudden a Pidgeon-House happens to be gutted and clear'd: No Satisfaction must be expected in that Case; the Pidgeon that flies into a strange Pidgeon-House, is accounted a wild Pidgeon. Those Pidgeon-Hunters are called *Kester-baze*, and *Kester perron*, i.e. Deceivers and Stealers of Pidgeons. Those Words, in a moral Sense, are scandalous, denoting an idle Fellow and a Pick-pocket: Truly the Pidgeon-Stealers spend whole days at that Trade, and are not diverted from it by the Severity of the Winter-Season" (Chardin 1927, 177). Fun no doubt for the Deceiver, but infuriating to the owner of the deserted tower who would feel as bereft as a beekeeper whose hive has swarmed.

Distribution of towers

By far the greatest number of towers survive within the oasis of Isfahan. The most distant seen was a group of four in a village ninety three kilometres to the southwest of the city (east of the main Shiraz road, 10 km beyond Shahrezah). Morier recorded "an immense number", ruined and in use at Komenshah, renamed Shahrezah (Morier 1818, 155). To the northwest, the occasional tower survives about a hundred kilometres from Isfahan. The only other group seen within the present-day borders of Iran are the rectangular towers described in Gulpayegan and Khunsar. Compared with those of Isfahan, the numbers here are very small.

The Afghan examples, near Herat, are in country which was part of Persia until two centuries ago. It is on the opposite edge of the plateau to Isfahan in another fertile river valley near what was another royal capital.

The absence of pigeon towers elsewhere is at first sight very surprising but it can probably be accounted for by the fact that the squabs were not used for food. A capital city produces a concentration of wealth which makes it profitable to intensify the agricultural output of a region which serves it; a steady demand for the luxury of melons over a long season would necessitate heavy use of fertilisers. Their production was the *raison d'être* of the pigeon tower which might today be termed a fertiliser collection centre (it was not, as said already, until after Isfahan ceased to be the capital that many of the towers began to fall to ruin.). Elsewhere, away from the capital, the individual land-owners and peasants could grow enough melons in season without going to the expense and trouble of building a pigeon tower in which to collect the relatively small amounts of fertiliser that might be needed.

The other use for the dung, for the manufacture of gunpowder, must not be forgotten. This too would lead to the siting of towers within easy reach of a capital city where an important part of the Shah's army would be based.

The success of a pigeon tower must also depend on the food available to the pigeons foraging for themselves on both crops and on uncultivated land. A rich oasis like Isfahan or a river valley like that of the Hari Rud would supply both, something which could not be claimed by many places on the Plateau. Nor could many regions supply sufficient water; melon crops depend on an exceptionally large supply for their irrigation.

It was the rare juxtaposition of these diverse factors which made the building of pigeon-towers worthwhile: demand for melons over a long season at up-market prices had to be centred in a region where the birds could successfully forage and the fruit be generously irrigated.

BIBLIOGRAPHY

Admiralty: Naval Intelligence Division, *Persia*, London 1945.

Australian Government Publishing Service, Experimental Building Station, *Notes on the Science of Building*, Canberra.

13 Earth Wall Construction, Revised 1971.

18 Pisé (Rammed Earth), Revised 1964.

20 Adobe (Puddled Earth), 1964.

22 Stabilised Earth, 1964.

Avitsur, S., *On the History of the Exploitation of Water Power in Eretz-Israel*, Tel-Aviv 1960.

Badawy, A., *Le dessin architectural chez les anciens Egyptiens*, Cairo 1948.

Bahadori, Mehdi, N., "Passive Cooling Systems in Iranian Architecture," *Scientific American* 238, No. 2, New York 1978.

Ball, W. and Hutt, A., *Persian Landscape*, London 1978.

Beaton, D., Letter to the *Gardeners' Chronicle*, January 13 1849.

Beaumont, Peter, "Water resource development in Iran", *Geographical Journal* 140, London 1974.

Beaumont, Peter, "Threat to Rural Iran", *Geographical Magazine* XLIX, No. 12, London 1977.

Beazley, E., "Greek Mills in Iran." *Architectural Review* CXXXVII, London 1967.

Beazley, E., "The Pigeon Towers of Isfahan," *Iran* IV, 1966.

Beckett, P., "Qanats around Kirman," *Royal Central Asian Society Journal* 40, London 1953.

Beckett, P., "Waters of Persia", *Geographical Magazine* XXIV, No. 5 London 1951.

Bellew, H.W., *From the Indus to the Tigris*, London 1874.

Bertram, C. & M.,"Pigeon Towers of Isfahan," *Geographical Magazine* XLI, No. 7, London 1969.

Binning, R.B.M., *A Journal of Two Years Travel in Persia*, vol. ii, London 1857.

Bonine, M.E., "From *Qanat* to *Kort*: Traditional Irrigation Terminology and Practices in Central Iran", *Iran* XX, 1982.

Boyce, M., "The Zoroastrian Houses of Yazd," *Iran and Islam*, Edinburgh 1971.

Calvert, N.G., "On Water Mills in Central Crete," *Transactions of the Newcomen Society* XLV, 1973.

Cambridge History of Islam, Cambridge 1970.

Cambridge History of Iran, Vol. I, Cambridge 1968.

Chardin, J., *Sir John Chardin's Travels in Persia*, London 1724, reprint 1927.

Cloudesley-Thompson, J.L., *Man and the Biology of the Arid Zones*, London 1977.

Coles, A and Jackson P., "A windtower house in Dubai," *Art and Architecture Research Papers*, London 1975.

Curwen, E.C., "More about Querns," *Antiquity* XV, 1941.

Curzon, G.N., *Persia and the Persian Question*, 2 vols, London 1892.

Denyer, S., *African Traditional Architecture*, London 1978.

Development Workshop, *The Caspian* , 1978; *Oman*, 1975. (Chichester, England).

Dickson, W.E.R., *East Persia*, London 1924.

Dieulafoy, J., *La Perse, la Chaldée et la Susiane*, Paris 1887.

Donaldson, B., *The Wild Rue. A Study of Muhammadan Magic and Folklore in Iran*, London 1938.

Edelberg, L., and Jones, S., *Nuristan*, Graz 1979.

Emanuel, W.V., *The Wild Asses*, London 1939.

English, P.W., "The Origin and Spread of Qanats in the Old World," *Proceedings of the American Philosophical Society* Vol. 112, No. 3, Philadelphia 1968.

Fathy, H., *The Arab House in the Urban Setting: Past, Present and Future*, London 1970.

Fathy, H., *Architecture for the Poor*, Chicago 1973.

Ferdinand, K., "Rice – Aspects of Cultivation and Treatment in East Afghanistan", *KUML*, Aarhus 1959.

Ferdinand, K., "The Horizontal Windmills of Western Afghanistan", *FOLK*, Copenhagen 1963 and 1966/7.

Filmer, H., *The Pageant of Persia*, London 1937.
Floyer, E.A., *Unexplored Baluchistan*, London 1882.
Folk-Edgar, G., *Textbook of Environmental Physiology*, 1974.
Forbes, R.J., *Studies in Ancient Technology*, Vols. I, II and VI, Leiden 1955.
Forbes-Leith, F.A.C., *Checkmate*, New York 1927.
Forster, G., *A Journey from Bengal to England*, London 1808.
Freese, S., *Windmills and Millwrighting*, Cambridge 1957.
Fryer, J., *A New Account of East India and Persia being Nine Years Travels 1672/81*, London 1698.
Fryer, J., ed. W. Crooke, *A New Account of East India and Persia being Nine years Travels 1672/81*, Vols. II and III, London 1912 & 1915.
Gabriel, A., *Durch Persiens Wüsten*, Stuttgart 1935.
Gabriel, A., *Aus den Einsamkeiten Irans*, Stuttgart 1939.
Gabriel, A., *Die Erforschung Persiens*, Vienna 1952.
Ghirshman, R., *Iran*, Paris 1951.
Ghiyath al-Din Sistani, *Ihya al-Moluk*, (Persian only),Tehran 1966.
Gibb, H., trans., *Travels of Ibn Battuta*, Cambridge 1971.
Goblot, H., *Les Qanats*, Paris 1979.
Gold, M., trans., *Tarikh-e Sistan*, Rome 1976.
Griffiths, J.F., *Climate and the Environment*, London 1976.
Grötzbach, E., "Anardarrah", *Afghanistan Journal* Vol. IV, Graz 1974.
Harrison, R.,"Traditional Monolithic Earth Walling Methods in Britain," *SPAB News*, July 1980.
Hedin, S., *Overland to India*, London 1910.
Herbert, T., ed. Sir William Foster, *Travels in Persia 1627-29*, London 1928.
Hill, D.R. trans., *The Book of Knowledge of Ingenious Mechanical Devices*, Dordrecht 1974.
Holford, I., *The Guinness Book of Weather, Facts and Features*, London 1977.
Housego, J., "The Southern Route to Mashhad: Caravanserais between Yazd and Torbat Haidariyeh", *Art and Architecture Research Papers* 6, London 1974.
Houtum-Schindler, A., *Reisen im Südlichen Persien 1879*, Berlin 1881.
Houtum-Schindler, A., *Eastern Persian Irak*, London 1898.
Humlum, J., *La Géographie de l'Afghanistan*, Copenhagen 1959.
Humlum, J., "Underjordiske Vandingskanaler," *Kulturgeografi* no. 90, December 1964.
Ibrahim, E. & D. Johnson-Davies, trans., *An-Nawawis Forty Hadith*, 1976.
Kaempfer, E., *Amoenitatum Exoticarum*, Lemgoviae (Lemgo) 1712.
Koenigsberger, O.H., *Manual of Tropical Housing and Building. Part One: Climatic Design*, London 1973.
Kozmin, P.A., *Flour Milling*, London 1917.
Lal, Mohan. *Journal of a Tour through the Panjab, Afghanistan, Turkistan, Khorasan and Part of Persia in Company with Lieut. Burnes and Dr. Gerard*, Calcutta 1834.
Lambton, A.K.S., *Landlord and Peasant in Persia*. London 1953.
Lestrange, G., *Lands of the Eastern Caliphate*, Cambridge 1905.
Lestrange, G., "Description of the Province of Fars." *Journal of the Royal Asiatic Society*, 1912.
Lewcock, R. and Freeth, Z., "Traditional Architecture in Kuwait and the Northern Gulf," *Art and Architecture Research Papers*, London 1978.
Locke, G., "Icehouses", *National Trust News Letter* No. 24 (Autumn 1975).
Lockhart, L., *Famous Cities of Iran*, Brentford 1939.
Lovell, A., trans., *Travels into the Levant by Monsieur de Thevênot*, London 1687.
Lunt, M.G., *Stabilised Soil Blocks for Building*, Overseas Building Notes No. 184, HMSO, London 1980.
Malcolm, N., *Five Years in a Persian Town*, London 1905.
Mellaart, J., *Çatal Hüyük, A Neolithic Town in Anatolia*, London 1967.
Merritt-Hawkes, O.A., *Persia, Romance and Reality*, London 1935.
Mez, A., *The Renaissance of Islam*, Patna 1937.
Michaud, R. and S., *Afghanistan*, Paris/London 1980.
Michell, G., ed., *Architecture of the Islamic World*, London 1978.
Morier, J., *A Journey through Persia ... in 1808 and 1809*, London 1812.
Morier, J., *A Second Journey through Persia, Armenia and Asia Minor to Constantinople between the Years 1810 and 1816*, London 1818.

Morier, J., *The adventures of Hajji Baba in Ispahan*, London 1824, Illustrated Reprint 1897.
Muir, W., *The Caliphate, its Rise, Decline and Fall*, Edinburgh 1924.
Myrdal, J. and Kessle, G., *Gates to Asia*, London 1972.
Naderi, B., "The Windmills of Khaf" in *Honar-e-Mardom* No. 177-8, (Persian only), Tehran 1978.
Najafi, Najmeh with Helen Hinkley, *Reveille for a Persian Village*, London 1959.
Napier, G.C., *Collection of Journals and Reports*, London 1876.
Needham, J., *Science and Civilisation in China*, Vol. 4, part ii, Cambridge 1965.
Noel, E., "Qanats", *Royal Central Asian Society Journal* Vol. 31, pt. 2, 1944.
Norton, J., *Report Study C89, Iran*, Unesco, Paris 1980.
Norton, J., *Report Study C90, Iran*, Unesco, Paris 1980.
O'Donovan, E., *The Merv Oasis*, Vols I and II, London 1882.
O'Kane, B., "Madrasa al Ghiyasiyya at Khargird", *Iran* XIV, 1976.
Olearius, A., *The Voyages and Travels of Ambassadors sent by Frederick Duke of Holstein to the Great Duke of Muscovy and the King of Persia*, London 1669.
Parkes, F., *Wanderings of a Pilgrim in Search of the Picturesque*, Vol. I, London 1852.
Polo, M., *The Travels*, Penguin Classics, London 1958.
Pope, A., & Phyllis Ackerman, *A survey of Persian Art*, Vol. III, Oxford 1938.
Redard, G., *Persien*, Zürich 1966.
Reynolds, J., *Windmills and Watermills*, London 1970.
Sackville-West, V., "Persian Gardens", in *The Legacy of Persia*, ed. A.J. Arberry, Oxford 1953.
Savage-Landor, A.H., *Across Coveted Lands*, London 1902, New York 1903.
Siroux, M., *Caravansérails d'Iran et petites constructions routières*, Cairo 1949.
Simpson, W., "Mud, a Material in Persian & Eastern Architecture," *Journal of the Society of Arts* XL, 1893.
Smith, A., *Blind White Fish in Persia*, London 1953.
Stevens, R., *The Land of the Great Sophy*, London 1962.
Stewart, C.E., *Through Persia in Disguise*, London 1911.
Stronach, D., *Pasargadae*, Oxford 1978.
Suffield-Jones, N., "Dovecotes and Gunpowder", *Surrey Archaeological Collections* Vol. 63, 1966.
Sulman, F.G., *Health, Weather and Climate*, Basel 1976.
Sykes, P.M., *Ten Thousand Miles in Persia*, London 1902.
Sykes, P.M., *Report on the Agriculture of Khorasan*, Simla 1910.
Tate, G.P., *Memoir on the Ruins of Seistan*, Calcutta 1910.
Tavernier, J.B., trans. J. Phillips, *The Six Voyages of Jean Baptiste Tavernier ... Finished in the year 1670*, London 1678.
Tavassoli, M., *Architecture in the Hot Arid Zone*, privately printed, Tehran 1975.
Wailes, R., "Horizontal Windmills", *Transactions of the Newcomen Society* XL, London 1968.
Wiedemann, E., *Aufsätze zur Arabischen Wissenschaftsgeschichte*, Vol. I, Hildesheim 1970.
Wills, C.J., *The Land of the Lion and the Sun*, London 1891.
Wotton, H., Reliquiae Wottonianae, *Of the Elements of Architecture*, London 1651.
Wright, S., "Place and Face: of Women in Doshman Ziari, Iran", *Women and Space*, ed. Ardener, S., London 1981.
Wulff, H.E., *The Traditional Crafts of Persia*, Cambridge, Mass. 1966.
Wulff, H.E.,"The Qanats of Iran," *Scientific American* 218 (4), 1968.
Yate, C.E., *Khurasan and Sistan*, London 1900.
Zestovsky, P.I., "Esquisses d'Architecture Afghane," *Afghanistan* IV, no. 3, 1949.

INDEX

(Arabic numbers refer to pages and italic ones to figures)